WHAT IF EINSTEIN WAS WRONG?

WHAT IF EINSTEIN WAS WRONG?

Asking the big questions about physics

Editor
BRIAN CLEGG

Foreword by
JIM AL-KHALILI

Ivy Press

First published in Great Britain in 2013 by
Ivy Press
210 High Street
Lewes
East Sussex BN7 2NS
www.ivypress.co.uk

This paperback edition published in 2015 by
Ivy Press

British Library Cataloguing-in-Publication Data
A CIP catalogue record for this book is available from the British Library.

ISBN: 978-1-78240-250-3

This book was conceived, designed and produced by **Ivy Press**

Creative Director: Peter Bridgewater
Publisher: Jason Hook
Art Director: Michael Whitehead
Editorial Director: Caroline Earle
Project Editor: Jamie Pumfery
Design: JC Lanaway
Illustrator: Ivan Hissey
Historical Text: Brian Clegg and Simon Flynn

Printed in China

Colour Origination by Ivy Press Reprographics

10 9 8 7 6 5 4 3 2 1

Distributed worldwide (except North America) by Thames & Hudson Ltd., 181A High Holborn, London WC1V 7QX, United Kingdom

Contents

FOREWORD

I have taught an introductory course on Einstein's theory of relativity to first year students at the University of Surrey for many years now. Just how many years became apparent when I recently bumped into an ex-student of mine who is now in her mid-thirties, she has a family and successful career as an academic physicist herself and who told me she remembers taking my class... seventeen years ago! Not only did I suddenly feel rather old, but I realised it was high time I passed the course on to someone else in my department. This I promise to do, soon.

In this course, I carefully go through the implications of Einstein's postulate that the speed of light is the same for all observers no matter how fast they are moving relative to each other. The students soon realise that this has some counter-intuitive implications, such as the idea that time itself flows at different rates in different frames of reference. I discuss the idea of four-dimensional space-time, go on to derive Einstein's famous equation, $E=mc^2$, from first principles and show how things start to look very different when we approach the speed of light. To students embarking on a physics degree this is great stuff and, so they assure me, far more interesting than their other courses – on optics, thermodynamics, advanced calculus or trying to get sensible data from experiments that never seem to work.

The highlight of my course is always the statement that nothing can travel faster than the speed of light. They ask 'but what if something could?' And 'what is so special about the speed of light anyway?' Or 'what makes physicists so sure that such a universal speed limit exists? Maybe we just haven't been imaginative enough'. I say, 'ok, what if an object could travel faster than light – what are the implications?' I proceed to carefully show how this leads to the inevitable conclusion that the object would have to move backwards in time in some reference frames, which is impossible since that would lead to a paradox.

This is in fact how we check many of the ideas and theories in physics; we subject them to the 'what if…' tests and push them to their limits. It is one of the most important tools we have to learn about the Universe and is often the way science, and physics in particular, progresses; we imagine the unimaginable scenario and see what our theory predicts about it. Sometimes, what we find strengthens our trust in a theory and sometimes it highlights a shortcoming or flaw in our thinking that leads to a breakthrough in understanding. This is when we must replace that theory with a better one. To the non-scientist, I think this highlights the sheer creativity of the process of science that is required to push our understanding forward. And it is such tremendous fun too.

J. S. Al-Khalili

Jim Al-Khalili

INTRODUCTION

Ask anyone to name a great physicist – or a genius – and the most popular answer is likely to be Albert Einstein. Yet even Einstein managed to get it wrong when it came to the more exotic aspects of physics. Though he was a major contributor to the foundation of quantum theory, he believed it was fatally flawed. He detested the way that it put probability at the heart of reality. According to quantum physics there was, for instance, no absolute value for a particle's location before it was measured – until then, it existed as a collection of probabilities. Einstein was convinced that underlying everything were hidden, real values. It was this that made him write to his friend Max Born: 'I find the idea quite intolerable that an electron exposed to radiation should choose of its own free will, not only its moment to jump off, but also its direction. In that case, I would rather be a cobbler, or even an employee in a gaming house, than a physicist.' Yet it was Einstein that was wrong, not the theory. There are no hidden values. Quantum physics really is that strange.

Einstein also made an error in describing the nature of the universe. When his equations for general relativity predicted that the universe should be unstable and contract or expand he put in a 'fudge factor' called the cosmological constant to keep the predictions in the unchanging condition he assumed represented reality. Yet soon after, Edwin Hubble discovered that the universe was expanding. The constant was not needed and Einstein would refer to it as his 'greatest mistake' – though in the fullness of time the equations would need a shift in the opposite direction to cope with the acceleration of expansion driven by the mysterious dark energy.

It shouldn't be too much of a surprise that Einstein could get it wrong, because science is not about absolute truth – it is our best understanding given the current data. Each section of *What if Einstein was Wrong?* includes a historical example that was once challenging but now is universally accepted. For the other entries we explore aspects of physics that still seem as strange as when they fooled great thinkers or that remain speculative. Each of our 'What If' answers is accompanied by a short 'What Then?' reflecting on the implications of the 'What If' being true, and some eye-opening 'What Gives?' facts and figures.

Albert Einstein made many contributions to physics, but his biggest impact came through laying the foundations of **Quantum Physics** and pulling together the theories that cover **Relativity and Time Travel**, which is why these are the opening two sections to this book. Although Einstein detested the strange probabilistic nature of quantum theory, his Nobel Prize winning paper on the photoelectric effect established that the quantum of energy provided by a photon of light was a real entity, making possible the discovery of the structure of the atom and the first steps down the road of understanding the weird behaviour of particles at the quantum level. Einstein's special relativity explored the relationship between space and time, while his general relativity provided the ultimate vindication of Newton's theory of gravity, showing how massive bodies warp space and time. Between them, these theories have provided the basis for the real theories of time travel.

Quantum theory tells us about the nature of matter and light, but it falls to **Particle Physics** to get the zoo of tiny particles that make up our universe in order, with our best current theory called the Standard Model. Recent developments at the Large Hadron Collider have seen some confirmation of the possible existence of the most mysterious part of this model, the Higgs boson – yet particle physics remains on the very edge of scientific knowledge with as much speculation as certainty, something that also applies to the most wide-ranging division of physics, **Cosmology**.

If particle physics and quantum physics are dedicated to the limits of smallness, cosmology goes to the other extreme and takes in the whole universe. Yet despite working on such enormous scales, and taking in events that we can only consider very indirectly like the big bang, cosmology

INTRODUCTION

is strongly linked to the world of the very small as quantum effects become highly significant in the early moments of the universe. Quantum physics also has an impact on the more detailed **Astrophysics**, which explains the nature and life cycles of stars and their strange cousins neutron stars and the enigmatic black holes. Although the black holes thought to be at the centre of our and other galaxies are vastly larger than the Sun, quantum theory is essential to understand their interaction with the universe around them – as is general relativity, which first predicted their existence.

It might seem that the surprises and delights in physics all originated from the overthrow of old ideas that occurred in the first half of the 20th century, but **Classical Physics**, the picture that held sway before quantum theory and relativity took hold, still provides its surprises, whether emerging from the deceptively simple laws of thermodynamics or even as old a challenge as working out why a mirror seems to present an image that is reversed left and right. Classical physics is usually more down to earth than the more modern aspects, but both classical and the new physics have made our modern **Technology** possible and it seems fitting to finish this exploration of the mind-bending aspects of physics with the most immediate aspect of the science – how it applies to our everyday life, which can still surprise and entertain us.

Physics is sometimes represented as a dry, mechanical science – and the way it is taught in schools it certainly can be. But at the same time it is the science that can best instil a sense of wonder, that can best provide excitement and amazement. Even Albert Einstein could occasionally be wrong-footed by physics. But he would not have seen this as a bad thing. Scientists like to be surprised. And physics can do this like nothing else.

QUANTU
PHYS

QUANTUM PHYSICS

INTRODUCTION
QUANTUM PHYSICS

When physicist Max Planck attended the University of Munich towards the end of the 19th century he was uncertain whether to opt for a career in physics or to build on his skills as a concert pianist. His professor, Philipp von Jolly, told Planck that physics was a dead-end pursuit. Apart from a couple of minor details, he explained, the physical theories of the day were complete. There was nothing left to do but add a few decimal places to experimental values and polish up the presentation.

Planck ignored his professor and went into physics – only to discover that those minor details would blow apart everything his 19th-century predecessors had assumed to be true. Planck showed that the only way to make sense of the interaction between matter and light was to view light as coming in little packets – quanta – rather than the continuous waves that everyone had assumed. And that apparently small change in viewpoint brought most accepted physics into question.

Quantum theory showed that the world of tiny particles that made everything up – from photons of light to the electrons, protons and neutrons of matter – behaved differently from the 'macro' world of everyday objects. We might imagine, for example, an electron orbiting an atom was like a satellite orbiting the Earth – but in reality the way electrons behave is much more strange and wonderful.

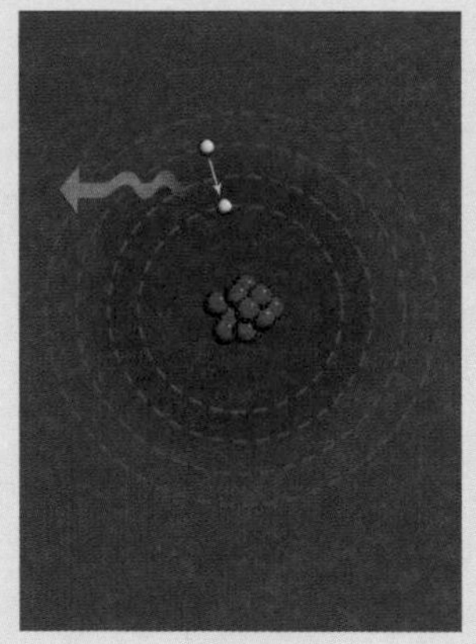

Richard Feynman, one of the greatest scientists to work on quantum physics, once pointed out that quantum theory described nature as absurd from the point of view of common sense – yet the theory agrees fully with experiment. 'Please don't turn yourself off,' he wrote, 'because you can't believe Nature is so strange. Just hear me out, and I hope you'll be as delighted as I am when we're through.'

Quantum physics brought us the idea of the quantum leap – the tiny change in the energy level and quantum state of an electron orbiting an atom. It also introduced the uncertainty principle, which elegantly shows that the more you know about one aspect of a quantum particle, the less you will know about another. And it brought a detailed mathematical understanding of the fundamental particles that showed, as Feynman had suggested, that common sense had no role to play.

Although much you will read in this chapter seems bizarre, quantum physics is not just an interesting theory. Without quantum effects, the Sun would not shine. Without quantum physics, light would not interact with matter the way it does, nor would atoms be stable. Without quantum behaviour, we would not have electronics, lasers or superconductors. Welcome to the hidden world of the very small.

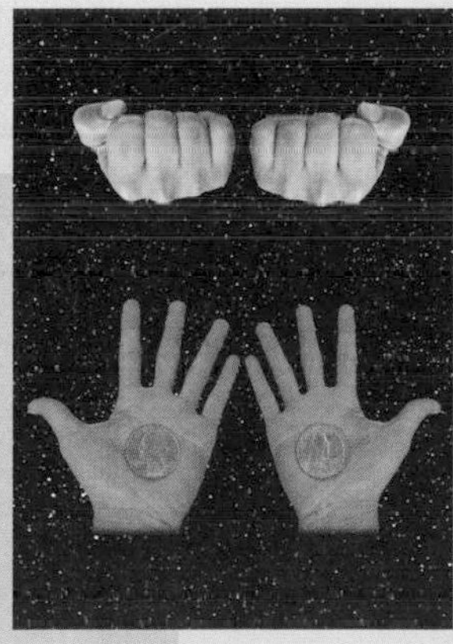

WHAT IF YOU MADE A QUANTUM LEAP?

Brian Clegg

We all know what a politician means when she says 'We've made a quantum leap forwards.' The leap is big, bold and important. But as a metaphor it's hard to imagine anything less suitable, because a real quantum leap is tiny in scale. The concept emerged from the early development of quantum theory and our understanding of what the structure of an atom is like. In the early years of the 20th century there was considerable doubt over whether atoms existed at all as physical objects, but as experimental evidence made their existence more and more likely, effort was focused on how they were put together. Experiments showed that atoms could emit the newly discovered electron particle, which was negatively charged – so the challenge was to devise a model that would allow electrons to combine with something positive to give an overall neutral atom. At first scientists thought that all the mass in an atom came from its electrons, so they assumed that even the simplest atom, hydrogen, contained a large number of electrons embedded in a massless positive matrix – the so-called plum pudding model. This picture was shattered when positively charged alpha particles were found to bounce off some atoms, showing that most of the mass was concentrated in the centre in a tiny, positively charged nucleus. Danish physicist Niels Bohr came up with a model of the atom similar to the solar system – light, negatively charged electrons orbiting a positively charged nucleus. The only problem with this is that orbiting electrons would give off energy in the form of light and spiral inwards. Bohr solved this problem by putting the electrons in his model on fixed orbits, like tracks: they could not exist between these orbits, but had to jump from one to another, making a quantum leap – the smallest possible change in energy they could undergo.

What Then?

Bohr's quantum leaps fitted elegantly with the discovery that atoms could only absorb light in chunks ('quanta'). It also explained how a spectroscope could identify the atomic components of a light-producing material, even in distant stars. The spectrum of light from a star contains black lines where light energies are missing; these energies correspond to quantum leaps in the matter the starlight interacts with and are used as a 'fingerprint' to identify chemical composition.

What Gives?

1,837 Number of electrons a hydrogen atom would need if all its mass really did come from its electrons. In fact a hydrogen atom has one electron.

2,000 Speed in km/sec at which even the relatively sluggish electrons of a small atom move around their nucleus. Electrons move so fast around the nucleus that relativity has to be brought into play to calculate their exact behaviour. (2,000 km/sec is equivalent to roughly 1,250 miles/sec.)

What Else?

What if we could see the atom? *See page 68*

What if atoms were not empty? *See page 72*

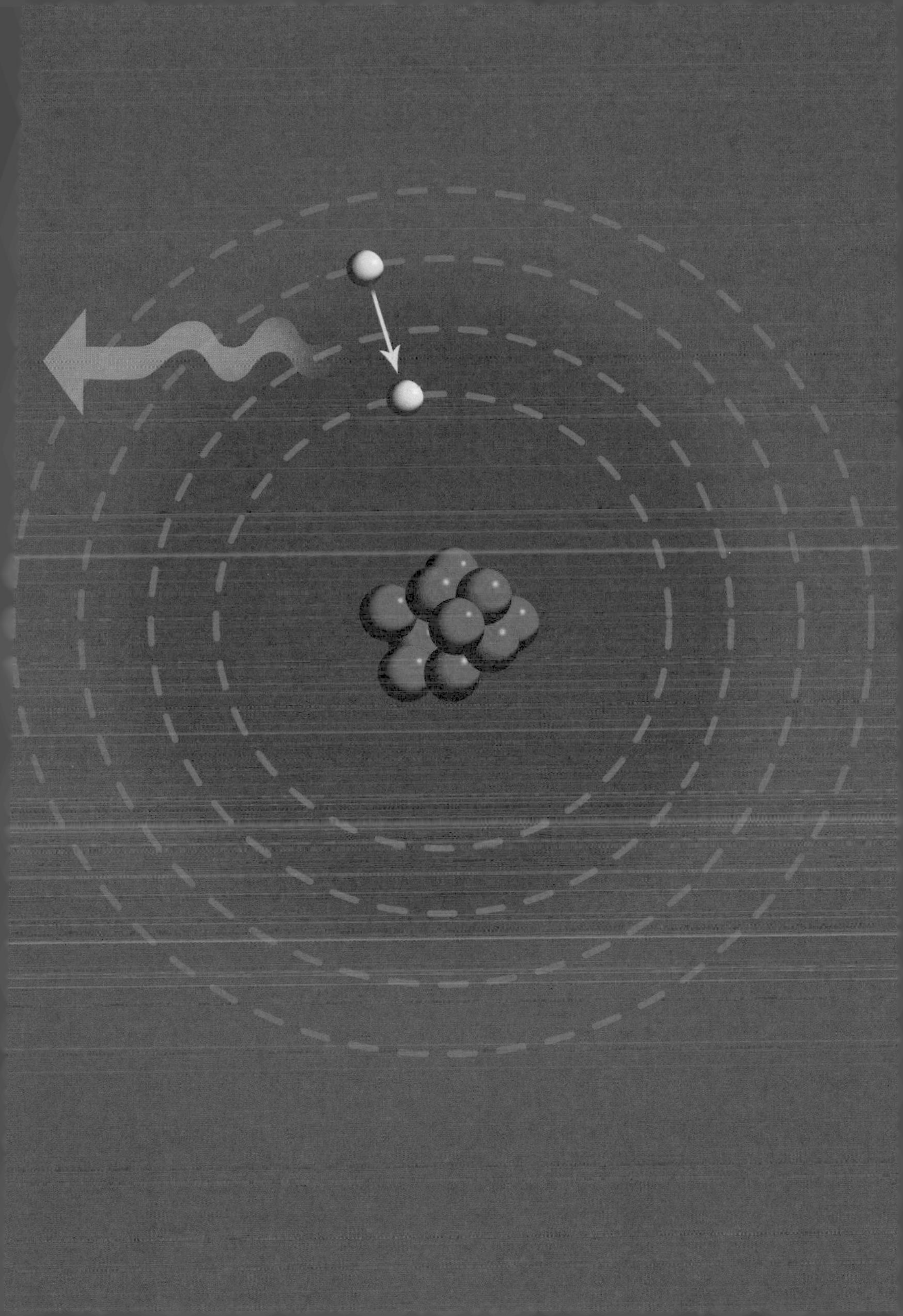

WHAT IF YOU COULD BE IN TWO PLACES AT ONCE?

Sophie Hebden

For an atom, space-time is inherently fuzzy. Quantum uncertainty becomes apparent at these small scales – making it impossible to know, with precision, both where an electron is and how fast it is moving. And tracking that electron down to make a measurement adds further uncertainty. If you put your electrons to the test to explore this quantum fuzziness, they change personality, depending on your experiment. Sometimes they behave like point-like particles, and sometimes like waves with long-range effects, enabling them effectively to be in more than one place at once. The first experiment to reveal that an electron can be in more than one place at once borrowed from the elegant two-slit experiment, first devised by English scientist Thomas Young in the 19th century to study light. If you shine light from a single source onto a pair of slits and onto a viewing screen, you will see an interference pattern of light and dark bands; this is because the light wavefronts spread out from the slits and interfere with one another, reinforcing some parts and cancelling out others, like concentric ripples in a puddle. Now ditch the light source and fire an electron at the slits, and it somehow interferes with itself on its way through, as if going through both slits at once. On the screen you can build up an interference pattern of light and dark bands – just like the experiment using light – by continuing to fire electrons singly through the slits. But watch out: if you check which slit your electron passes through, your interference pattern disappears and you get particle behaviour again. Two places at once? No problem for an electron. But how big can you go? By trying ever larger objects, these experiments probe the transition between the quantum and the everyday world. It could even work for living organisms. Perhaps the quantum world is not so far away after all.

What Then?

The wave behaviour of matter in computations can be described using a quantity called a wave function. From this you can calculate the likelihood of finding a particle somewhere at a given time. It's not until you make your measurement that the wave function collapses and the particle 'chooses' a particular location. The physical reality or not of the wave function and what happens during observation to make it collapse – called the measurement problem – are ongoing debates about quantum mechanics.

What Gives?

1927 The year in which American physicists Clinton Davisson and Lester Germer stumbled upon the wave nature of electrons when trying to undo a mistake during tests on nickel.

430 The largest molecules (in number of atoms) shown to act like waves. These molecules measure up to 6 nanometres across, the size of small viruses.

What Else?

What if Schrödinger lost his cat?
See page 22

What if light is not a wave?
See page 26

WHAT IF THE UNIVERSE WERE RANDOM?

Brian Clegg

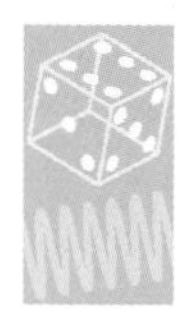

Every time you look out of a window at night you are undertaking an experiment that had English physicist Isaac Newton baffled and caused Albert Einstein no end of agonizing. Most of the light from the room will pass straight through the window. You can check this by going outside and looking in. But some of the light will be reflected back into the room, making the night-time window act like a mirror. Think of what is happening in terms of the individual photons – quantum particles of light – that hit the window. Some photons will reflect, others will pass through. But what decides the fate of any particular photon? Newton, who thought light was made up of particles called corpuscles, decided that particles must reflect if they hit imperfections in the glass – but polishing the surface doesn't make the reflection go away. In fact this is one of many behaviours of quantum particles that has randomness at its heart. There is no way of telling what a particular photon will do. We can work out the probability of a particular outcome, just as we can establish that the probability of getting a head with a coin toss is 50 per cent. But which specific outcome will occur for a particular photon is just as random as heads or tails when you toss a coin – more so, in fact, because you could predict a coin toss if you had all the physical data, whereas the behaviour of the photon is truly random. The same randomness applies to when a particular radioactive particle will decay and a whole host of other occurrences at the quantum level. This infuriated Einstein. He wrote to Hedwig Born, wife of physicist Max Born, 'I find the idea quite intolerable that an electron exposed to radiation should choose of its own free will, not only its moment to jump off, but also its direction. In that case, I would rather be a cobbler, or even an employee in a gaming house, than a physicist.' It was this aversion to randomness that inspired Einstein's famous comment 'God does not play dice.'

What Then?

Such was Einstein's distaste for this randomness that he came up with a series of thought experiments to prove quantum theory wrong – yet time and again, his opponents defeated his arguments. Eventually Einstein proved that either quantum theory was wrong and there was hidden information rather than randomness, or it should be possible for two particles to communicate instantly at any distance. This attempt to discredit the theory was the birth of the strangest aspect of quantum theory, entanglement (see page 28).

What Gives?

22 Percentage of light reflected back into a room by a typical double-glazed window.

1935 Year in which Einstein, with Russian-born physicist Boris Podolsky and American physicist Nathan Rosen, came up with the paper 'Can Quantum-Mechanical Description of Physical Reality Be Considered Complete', usually called EPR, attempting to dismiss quantum randomness.

What Else?

What if you made a quantum leap? *See page 16*

What if you could instantly signal the stars? *See page 24*

WHAT IF SCHRODINGER LOST HIS CAT?

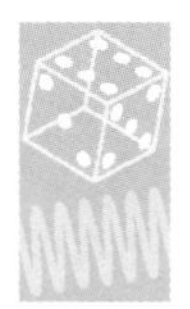

Danish physicist Niels Bohr, one of the founders of quantum theory, remarked that 'those who are not shocked when they first come across quantum theory cannot possibly have understood it'. One aspect of the theory that particularly took scientists aback was the impression that reality only resolved itself when it was observed. This was principally because probability, previously an anathema to physicists, was now understood to lie at the dark heart of subatomic events. About 150 years before the birth of quantum theory, French scientist Pierre-Simon Laplace had suggested that if the position and speed of every particle in the universe could be known at a single instant in time, their past or future could be calculated; in effect, we lived in a clockwork universe. But quantum theory revealed that electrons exhibited wave and particle properties and were an example of matter waves. In 1926, Austrian physicist Erwin Schrödinger advanced an equation that described matter waves. Schrödinger's equation showed them to be probability waves. The position of a particle could no longer be predicted but existed only as a probability. In a bid to pin down quantum theory, Bohr and his followers argued that this probability was resolved only when the particle was observed. Despite playing a key role in its formulation, Schrödinger abhorred this interpretation and proposed a thought experiment designed to show the ridiculousness of the situation. In it, a cat was put in a sealed steel box along with a small quantity of a radioactive substance. Over the course of an hour there was an equal chance of one of its atoms decaying or not: if it did, this would trigger the release of a poison, which would kill the cat. Until we look inside the box, the state of the cat exists only as a probability – therefore it is both alive and dead. For Schrödinger, this was nonsense. Ironically, a thought experiment designed to demonstrate the absurdity of quantum theory has turned out to be one of its greatest marketing pieces.

Simon Flynn

What Then?

Not all interpretations of quantum theory require observation to resolve reality. The many-worlds version by American physicist Hugh Everett III suggests that two realities exist once the box is opened, one in which the cat is observed to be alive and one in which it is seen to be dead. The two realities are completely separate, equally real and unable to interact with each other.

What Gives?

Quantum suicide Versions of the thought experiment that take the cat's point of view have been put forward.

2 Number of observers in another extended version of the thought experiment, known as 'Wigner's friend'. One observer conducts Schrödinger's experiment and another receives the information about its result.

What Else?

What if you could be in two places at once? *See page 18*

What if the universe were random? *See page 20*

What if light is not a wave? *See page 26*

WHAT IF YOU COULD INSTANTLY SIGNAL THE STARS?

Sophie Hebden

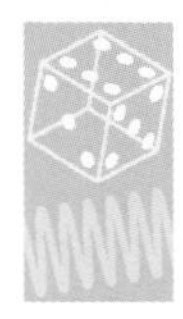

There's a simple game in which you hide a small object – say, a coin – in one hand, then present both of your hands as fists for your opponent to guess where it is. If the first guess reveals an empty palm, you automatically know the coin is in the other hand. Now imagine playing the game with two coins, one in each fist. Your opponent chooses a fist and you reveal the coin: perhaps it is heads up. Open your other fist and the second coin has a 50-50 chance of being tails, whatever the first result. How strange it would be if every time you checked your second coin it was the opposite of the first. That's exactly what happens when two quantum particles are entangled: checking the state of the first (say, a photon that you measure to be vertically polarized) automatically determines the state of the second (in this case, it would be horizontally polarized). And it doesn't matter how far apart the entangled particles are: they act as one. Entanglement was first predicted in 1935 by Austrian physicist Erwin Schrödinger as the solution to the problem of what happens when quantum particles interact and then separate. Einstein baulked at its 'spooky action at a distance', declaring that quantum mechanics must be incomplete – how can one particle's status be communicated instantly through space to another? But really there's no secret signal travelling faster than light: when we measure one particle in the entangled system, the possible states for the other particle reduce to one, so that we have a definite value for it, just as in our example of a coin in the fist – discovering an empty fist reveals that the coin is in the other hand. Nearly 80 years on, quantum entanglement is here to stay, and our experiments are becoming more and more sophisticated, involving many-particle systems, and stretching longer distances. Much of the research is directed at its exciting uses: entanglement underlies the technology for quantum computing and quantum cryptography.

What Then?

Entanglement is a superluminal effect (it travels faster than light), but this doesn't mean we can use it for superluminal signalling (sending information at speeds faster than light) because nothing is actually travelling. We can, though, use it for super-secure signalling, in which you beam entangled photons to a receiver. Any attempt secretly to observe the entangled signal would affect its state, making tampering obvious to the users. This underlies quantum key cryptography (using quantum effects to encode data), where distance entangled photons can be sent is the main challenge.

What Gives?

144 Longest distance (in km) for entanglement through free space. It was done between islands of the Canary Isles. (144 km = 89 miles.)

14 Record for the largest number of entangled particles in a quantum computer.

Beijing, London & Tokyo Cities that contain dedicated fibre-optic cables for sending secure quantum keys.

What Else?

What if Schrödinger lost his cat?
See page 22

What if you beamed me up?
See page 28

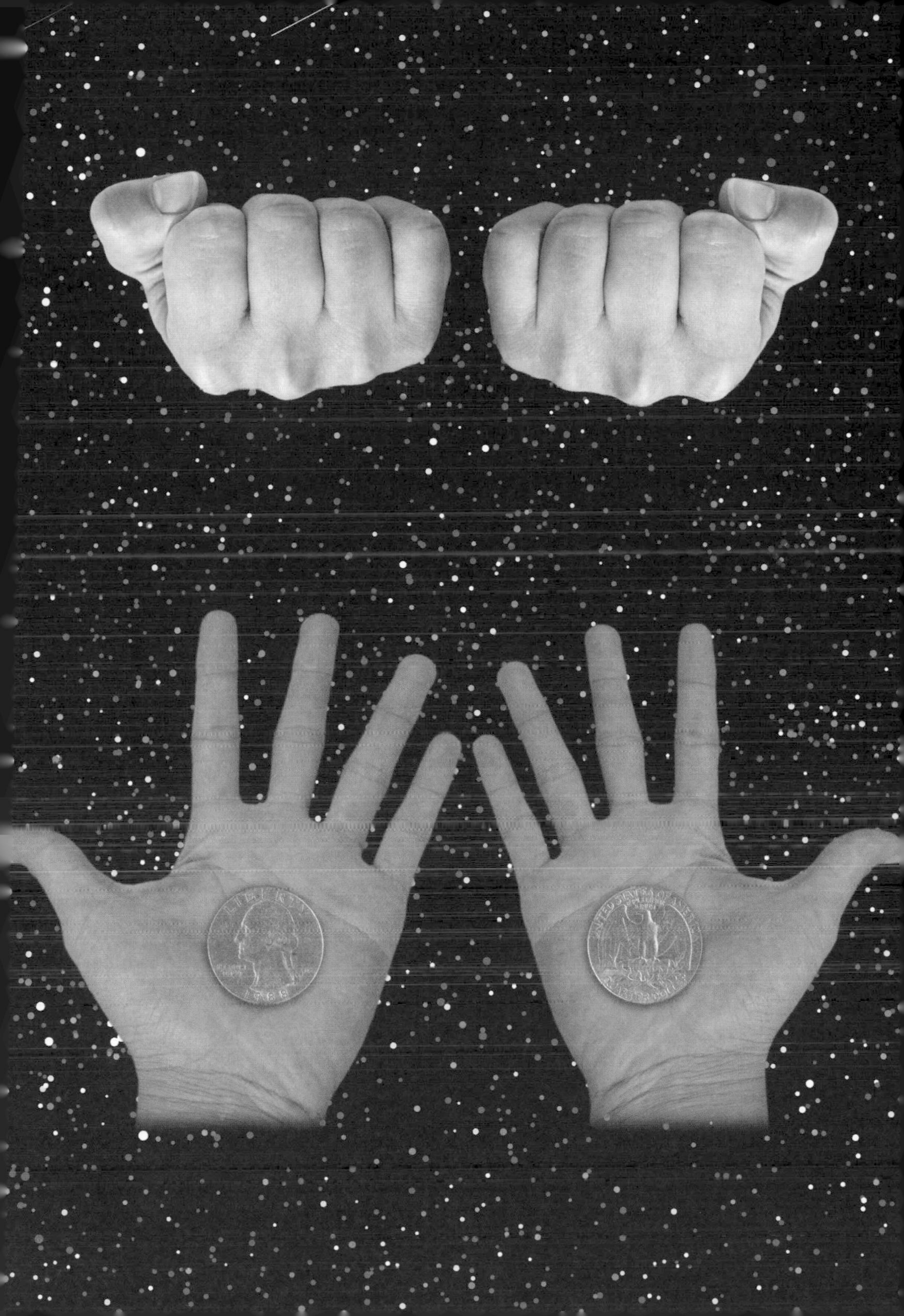

HISTORICAL
WHAT IF LIGHT IS NOT A WAVE?

In 1801 English scientist Thomas Young showed light to be a wave when he performed his double-slit interference experiment. It used an opaque material with two holes or slits cut into it, with a light source situated on one side and a screen on the other. When light was emitted by the light source, an alternating pattern of dark and bright spots was observed on the screen. Young's experiment demonstrated that when light waves overlap, they interfere with each other, producing a pattern in exactly the same way that water and sound waves do. The evidence couldn't have appeared more conclusive. But all was not what it seemed.

The roots of discontent started to appear in 1900 when German physicist Max Planck's theory of black-body radiation proposed that objects absorb or emit electromagnetic radiation (light) in multiples of very small, discrete units, which he called 'energy quanta' (quanta came from the Latin for 'how much'). Up until that point, an increase and decrease in energy had been thought of as being smooth and continuous, like an elevator moving up and down. Instead, it now appeared that energy jumped in tiny steps.

This provided a key insight into something that had been puzzling scientists during the previous 30 years – the photoelectric effect. Experiments had shown that when electromagnetic radiation was directed at a variety of metals, electrons were emitted from their surface. Within the effect were some very surprising associated results. Firstly, if the intensity of light changed, so did the number of electrons emitted – a lower intensity resulted in fewer electrons. But the energy of each emitted electron remained the same.

However, German physicist Philipp Lenard demonstrated in 1902 that changing the frequency (colour) of the light affected the energy of the individual emitted electrons – the higher the frequency the greater the energy. Nevertheless, when decreasing the frequency there would come a point at which no electrons were emitted. And what this minimum frequency, known as the threshold frequency, was depended on the metal in question. For example, caesium will emit electrons when yellow light is shone on it but platinum won't.

None of this made sense if light was a wave. For example, wave theory said that the greater the intensity of light the greater the amplitude of the wave – in which case, one would expect the kinetic energy of the emitted electrons to increase. And why would changing the frequency have the effect it did, especially in comparison to intensity? No matter how bright (intense) your light source, if the frequency was below the threshold then no electrons would be emitted.

The answer to the puzzle came in 1905 when Albert Einstein took Planck's idea of the quanta and applied it to the photoelectric effect. He suggested that light was made up of quanta of light energy, which we now call photons. And the energy of a photon depended on the frequency of the light. Increasing the intensity of the light would result in there being more photons but the energy of each photon would remain the same. Increasing the frequency, however, would also increase the amount of energy carried by each photon. This explains perfectly the photoelectric effect.

Einstein's explanation came the same year as his proof of the existence of the atom and his special theory of relativity. Yet it was for his work on the photoelectric effect that he was awarded the Nobel Prize for Physics in 1921.

WHAT IF YOU BEAMED ME UP?

Brian Clegg

Quantum entanglement – an interaction between two quantum particles in which, while remaining separate, they act as one – makes possible on a very small scale one of the most impressive devices of science fiction: the teleporter or matter transmitter. It provides a mechanism for sending a solid object from place to place without it passing through the space in between, travelling as pure information. When this idea was used for the transporters on the US TV show *Star Trek* it was simply there to save the special effects team the expense of simulating a shuttlecraft landing, and the writers were unaware that they seemed to be achieving the impossible. Everything from *Star Trek*'s Captain Kirk to a virus is made from a collection of quantum particles, and it is impossible to make a measurement of a quantum particle without changing it. This appears to prevent anything from making an exact copy of such a particle – this idea even has a name, the no cloning theorem. But entanglement provides a get-around to this limitation. We start with an entangled pair of particles, one at each end of your teleport. The first entangled particle is made to interact with the particle you want to transmit – call that one James. In the process, the 'James' particle is wiped of its identity and some information is produced that is sent to the receiver end of the teleportation device. There this information is used to decide what process to put the receiver's entangled particle through – and the result is to turn that into a clone of James. This is possible because we never tried to find out all the information about James – some of it travelled unseen down the entangled link. It might seem that because the spooky link of entanglement happens instantaneously, quantum teleportation could achieve the dream of sending an object faster than light, but because one part of the process involves transmitting information by conventional means, teleportation is limited to light speed or slower.

What Then?

If it ever became possible to use quantum teleportation to transport a human being, in *Star Trek* fashion, it would not be an attractive mode of transport. Teleportation does not move the component parts of an object from place to place, it makes a perfect copy and, in the process, destroys the original. The new you would be indistinguishable from the original. It would have your memories, your mind. But that wouldn't stop your body from being entirely disintegrated.

What Gives?

2004 Year in which Austrian physicist Anton Zeilinger and his team achieved teleportation of entangled photons across the River Danube, showing the process works over distance and away from lab conditions.

1 billion atoms a second Even scanning at this speed it would take 200 billion years to teleport every atom in a human being.

What Else?

What if you could instantly signal the stars? *See page 24*

What if we could compute with quanta? *See page 30*

WHAT IF WE COULD COMPUTE WITH QUANTA?

Brian Clegg

The computers that we encounter in our daily lives handle information broken down into the basic unit of the bit (a contraction of binary digit, either 0 or 1). The number of bits available, and the time it takes to access them, puts a limit on the potential power of any conventional computer. But quantum physics provides the opportunity to take the bit into a whole new dimension by making the fundamental unit of storage the state of a quantum particle, such as its spin. Measure the spin of a quantum particle in any particular direction and it will always come out either up or down. We can't predict what the value will be – before the measurement it exists in both directions at once. But we can discover the probability of it being up or down. It might be, say 60 per cent chance of up, 40 per cent down. If we can use the state of the quantum particle – called a qubit – as the equivalent of a bit, then instead of just holding a value of 0 or 1, the qubit holds this complex directional information, which can represent an infinitely long decimal value, giving the computer greatly enhanced capability. Computing with qubits is not easy. Many research teams are working on quantum computers, but as yet the new type of computer has only been able to handle relatively simple problems such as 'What are the factors of 15?' This reflects the difficulty of keeping qubits stable – most quantum computers currently only have a handful of qubits compared with the trillions of bits in the working memory of a typical home computer – but there is little doubt that they will eventually be scaled up. The other problem with a quantum computer is getting information in and out. The strange phenomenon of quantum entanglement, which enables particles to share quantum information without corrupting it, seems essential to making this possible.

What Then?

Even though we don't yet have useful quantum computers, we do have algorithms – mathematical recipes – that would run on them and enable a decent-sized quantum computer to outperform any conventional machine. One such algorithm makes it easy to find the prime number factors of a huge number (an ability that would make it possible to break most current computer encryption) while another, the so-called 'quantum needle in a haystack algorithm', makes it much quicker to search for information in messy data that has no structure or index.

What Gives?

1,000 Number of operations needed for a quantum computer using the 'quantum needle in a haystack algorithm' to make a search that would take a conventional computer up to 1 million tries (and would average 0.5 million tries).

2 Number of qubits in the first working quantum computer, which was capable of running simple algorithms.

What Else?

What if you could instantly signal the stars? *See page 24*

What if you beamed me up? *See page 28*

WHAT IF THERE WERE A SMALLEST DISTANCE?

Rhodri Evans

We are used to thinking of space and time as continuous – that is, we think that in theory we could measure ever smaller distances and intervals of time. But this may not be true. Space and time may be quantized (come in tiny, fixed-sized chunks), and there may be a smallest length and a smallest time interval in Nature. We call these the Planck length (or distance), denoted by l_P and the Planck time, denoted by t_P. Both are very, very small. The Planck length is based on a distance that is a combination of Planck's constant, the universal gravitational constant and the speed of light. (Planck's constant is an unchanging measure that describes the ratio of a photon's energy to its frequency. The universal gravitational constant is an unchanging measure that gives the ratio of force to mass and distance in the operation of the law of gravitation.) About $1.5x10^{-35}$ m (or 1.5 divided by 1 with 35 zeroes after it), the Planck length is the distance light travels in a Planck time. Compare this to the size of a proton – about $1x10^{-15}$ m. This makes the ratio of the Planck distance to the size of a proton about the same as the ratio of the size of a proton to 100 km (62 miles). The Planck time is produced by the same combination of Planck's constant, the universal gravitational constant and the speed of light, but for the Planck time we divide by the speed of light to the power 5, not 3 as is the case for the Planck distance. The Planck time is about $5.5x10^{-44}$ seconds. The ratio between the Planck time and one-billionth of a second is 10 million times less than the ratio between one-billionth of a second and the age of the universe. Our current belief is that at such small distances and times we need a quantum theory of space and time, not the classical theory we use. Scientists have been working on a quantum theory of gravity for over 50 years, but such a theory eludes us still. It means that we currently cannot understand the earliest moments of the big bang, when the time was less than the Planck time.

What Then?

Most theories of quantum gravity predict that space-time becomes 'foamy' at scales of the Planck length. In theory, two locations separated by 1 Planck length would be impossible to tell apart, no matter how closely we looked. Similarly, two events separated by 1 Planck time would seem to be simultaneous, no matter how accurately we were able to time the events.

What Gives?

Infinite Current calculation for the density at the centre of a black hole. Infinities in physics suggest a problem with the theory, so to understand the central parts of black holes we need to consider the Planck length, and hence need a quantum theory of gravity.

Zero Current theory for the size of the universe at its beginning. At this time its density was infinite. The Planck length makes the idea of zero distance impossible, so we need a quantum theory of space-time to understand the beginning of the universe properly.

What Else?

What if you could be in two places at once? *See page 18*

What if everything were made of string? *See page 84*

2

RELATIV
TIME T

RELATIVITY & TIME TRAVEL

INTRODUCTION
RELATIVITY & TIME TRAVEL

At its simplest, we can claim that relativity is nothing more than the observation that movement is relative. When we say something is moving at 100 m/sec (roughly 325 ft/sec), that only has meaning when compared with something – in everyday use, we would probably mean relative to the surface of the Earth. But for Albert Einstein this presented a problem when he thought about light.

Einstein's first venture into relativity, special relativity, was driven by the then new understanding of just what light was. Scottish physicist James Clerk Maxwell had shown in the 19th century that light was an interaction between magnetism and electricity. Maxwell explained how moving electricity could create moving magnetism that created moving electricity and so on indefinitely. The process could be self-supporting, hauling itself up by its own bootstraps. That's what light is, a self-supporting wave of electricity producing a wave of magnetism, which produces electricity and so on. But this only works if light moves at one speed in any particular substance – the speed of light. It is only by travelling at this particular speed that magnetism and electricity become self-supporting.

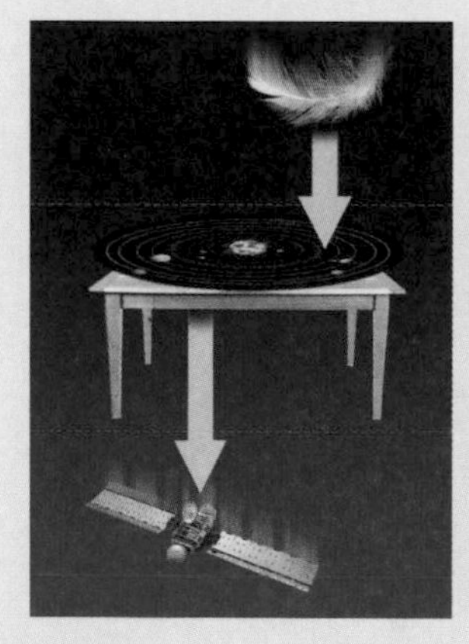

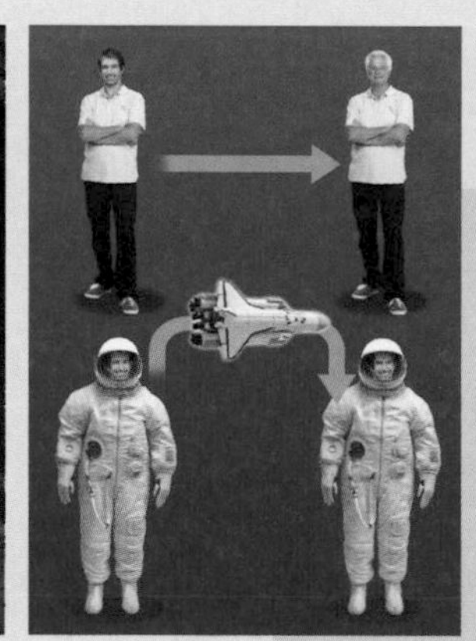

Einstein imagined flying alongside a beam of light at exactly the same speed. If he did, for him the light wouldn't be moving. But if it didn't move at just the right speed, it couldn't exist. The light would disappear. Whenever anything moved towards or away from a beam of light the speeds should combine, making light travel at the wrong velocity to exist, forcing it to wink out of existence. Einstein deduced that light must always move at the same speed, however you travel with respect to it.

When you plug this unique behaviour into the equations of motion, strange things begin to happen. To fix one thing, other aspects of reality have to vary. Once the speed of light is fixed, a moving object gets more mass, it squashes up in the direction of motion, and time slows down for it. This is special relativity. Einstein would later extend this to general relativity, which provides an explanation for the working of gravity as a warp in space and time.

Perhaps the most fascinating outcome of Einstein's theories of relativity is that it is possible to use them to theorize about travelling through time. Surprisingly, there is nothing in the laws of physics that prevents time travel. By moving very quickly, making time slow down for you, you will move into the future of those you left behind; general relativity provides various means to manipulate spacetime and theoretically reach the past. Stand by for an exciting ride.

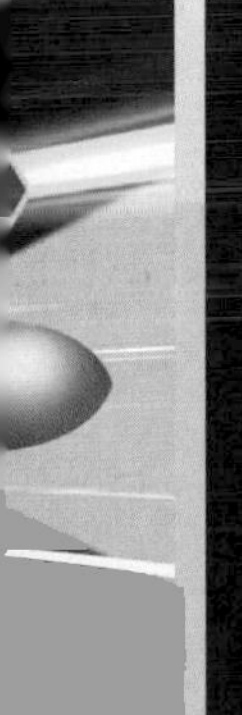

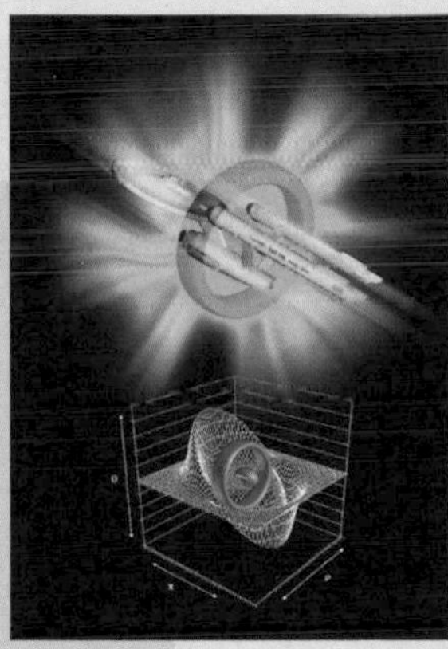

WHAT IF YOU WENT BACK TO THE FUTURE?

Rhodri Evans

Albert Einstein's special theory of relativity revolutionized our concept of space and time. It showed that time is relative, meaning that it can pass at different rates depending on your state of motion. As one approaches the speed of light, time starts to pass more slowly – we call this 'time dilation'. The effect is essentially negligible until one approaches about 75 per cent of the speed of light, but becomes more and more powerful as one gets closer to the light speed. If you were to travel at the speed of light itself, time would actually stand still for you. It is the effect of time dilation that makes time travel into the future possible. The 'twin paradox' illustrates how the effect works. One of a pair of twins goes on a space trip lasting 10 years according to his clocks and calendar. During this trip his spaceship travels at speeds approaching the speed of light. Time will pass more slowly for him than for his twin who stays on Earth, so when he returns to Earth he will be younger than his twin. If he were to travel at an appropriately high speed he could return to find that 40 years had passed on Earth, to find his twin now 30 years older than he is. This effect is not science fiction. We see it every day when particles called muons – created high in the Earth's atmosphere by high-energy cosmic rays – reach the ground. The muons should decay before they reach the ground, but they are travelling so fast that time passes more slowly for them, allowing them to reach the ground before they have decayed. Satellites orbiting the Earth in GPS systems have to take time dilation effects into account as they orbit the Earth at 3.1 km/sec (1.9 miles/sec) – more than 11,000 km/h (6,800 mph). If you could travel at close to the speed of light you could jump to the future in an instant, as the character Marty McFly does in the *Back to the Future* films.

What Then?

One remarkable consequence of time dilation is that if we were ever able to travel at speeds where its effects become important it would make travelling to other star systems possible. Although the nearest star system is 4.2 light years away, if we could travel at a speed where time were dilated by a factor of 4 (around 97 per cent of the speed of light) we could actually make this trip in just over one year – although on Earth over four years would have passed.

What Gives?

0 seconds Time needed for a photon's journey – of any length. For a photon travelling at the speed of light, time stands still and distances become zero. Thus, for a photon all trips are instantaneous.

3/100 of a second A proposed privately funded mission to Mars in 2018 is to take 501 days at around 3.5 km/sec (2.2 miles/sec): 12,600 km/h (7,800 mph). But even at ten times the speed, the time dilation effect would only amount to 3/100 of a second.

What Else?

What if you could journey into the past? *See page 40*

What if you hit warp speed? *See page 46*

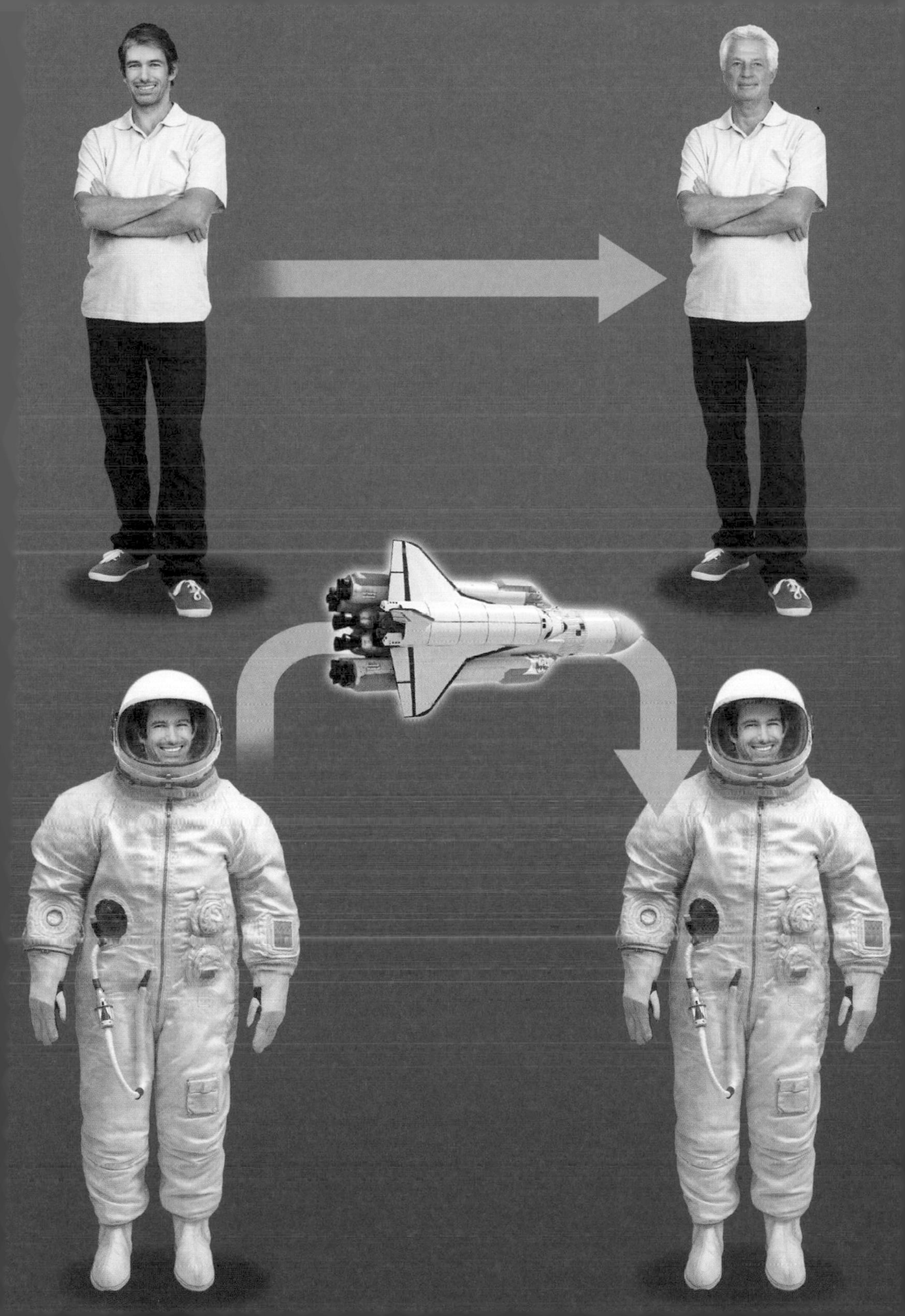

WHAT IF YOU COULD JOURNEY INTO THE PAST?

Brian Clegg

A theoretical physicist will tell you that travelling into the past is just a matter of engineering. All you need is to make a wormhole – a tear in reality that links two points in space-time – keep it open with negative energy and fly through it. The only problems with this are that wormholes are entirely theoretical, no one has ever seen one, we don't know how to make them and we certainly don't know how to travel through one safely. Or how to produce more than a tiny amount of negative energy. An alternative approach is to take a string of neutron stars, form them into a cylinder and spin the cylinder at high speed: fly around this cylinder and you've a time tunnel into the past. But again there are problems. We know where there are thousands of neutron stars, out in our galaxy. But to make a time machine you would need to cross hundreds of light years of space, grab yourself around 10 of these massive stars, haul them together and make a cylinder out of them. And then you've got somehow to stop the natural tendency of that cylinder to pull itself into a sphere and collapse as a black hole. Then spin it round extremely quickly. These are feats that are millions of years beyond today's technology, but the theory seems to be sound. However, even if we ever made such a device, we couldn't use it to visit the dinosaurs. That is because the furthest it can travel back is the point at which the machine was first switched on. Such devices don't reverse time, they jump to a place where time runs more slowly. Imagine I had a box in which time was virtually at a standstill. After a year of setting it up, the box's time would be a year behind mine. If I could jump into the box's world, I would travel back a year, but no further. These machines are much the same.

What Then?

If you can move into the past, time paradoxes emerge. What if, for example, you went into the past and killed one of your parents before you were born? How could you go back if you no longer existed? Or, what if I sent copies of the book you have in your hands to the contributing authors before it was written – and they simply copied out their sections? Who would have written the book? Some say the only explanation would be if travelling though time meant shifting to a parallel universe.

What Gives?

100 million tons

Weight of a piece of neutron star the size of a grape.

27,000 years

The time the *Apollo 10* spacecraft, our fastest vehicle to carry people so far, would take to cross a light year.

Frame dragging

An effect of general relativity by which a rotating mass pulls space and time with it, like a spoon rotating in honey. This is how the rotating cylinder works.

What Else?

What if you went back to the future? *See page 38*

What if there were parallel universes? *See page 80*

WHAT IF TIME RAN BACKWARDS?

Brian Clegg

It is hard to imagine time running backwards. Even the theoretical possibilities for time machines are built instead on the idea of jumping to somewhere where time passes more slowly. And yet physics is often indifferent to the direction in which time flows. Scottish physicist James Clerk Maxwell's equations for the interaction between electricity and magnetism – which explained the nature of light – had a strange implication. The equations had not one, but two solutions, each equally valid. One produced 'retarded waves' – light as we know it; the other, 'advanced waves', which started from the destination and ran backwards in time to reach the source. The response to this was to ignore the advanced waves because they didn't make sense – hardly a scientific approach. Advanced waves remained all but forgotten until American physicists John Wheeler and Richard Feynman were seeking to explain the way an atom behaves when it emits light. The atom recoils as the photon of light leaves – similar to the way in which a gun is pushed back when it fires a bullet. But the world of the atom is a complex place. The electromagnetic interaction involved in producing a recoil should result in a feedback loop. This 'self-interaction' would produce values that headed off to infinity. Wheeler and Feynman found it was possible to undo this unfortunate outburst of infinity by bringing in advanced waves. The life cycle of a photon typically involves it being produced by an electron undergoing a quantum leap in one atom. The photon then crosses space before being absorbed by an electron in another atom. The physicists suggest that two photons were involved, the second leaving the absorbing atom, moving backwards in time and arriving at the source at just the right moment to cause the recoil. There is no way of proving this picture from what is observed, but it makes sense of Maxwell's equations, explains the recoil and shows that time – at least for these photons – just may run backwards.

What Then?

If advanced waves exist, there is a theoretical way to use them to signal backwards in time. We understand that a photon is only emitted if it is going to be absorbed. So if we can identify a direction in space that has few absorbers, then push an absorber into place at a distant location in that direction, our action should be reflected back at the light source – the rate of photon emissions at the light source should rise before we put the absorber in place.

What Gives?

2 Number of photon releases involved in a single photon interaction between atoms, according to Wheeler and Feynman's theory: if the theory is true, a photon is actually two photons, each with half the total energy, each travelling in opposite directions in time and space.

10–20 Number of minutes it typically takes light to reach Earth from Mars. An advanced wave should leave Mars the same number of minutes before NASA transmits a signal from Earth.

What Else?

What if you could journey into the past? *See page 40*

What if you could tap into tachyons? *See page 44*

WHAT IF YOU COULD TAP INTO TACHYONS?

You won't see it on a road sign, but the speed limit of the universe is 299,792,458 m/sec (983,571,056 ft/sec). Nothing can travel faster than the speed of light in a vacuum – that includes radio signals, spaceships, subatomic particles or any sort of information. This speed limit doesn't need enforcement because, according to Einstein's theory of special relativity, it would take infinite energy to accelerate up to that speed. But what if a particle exists that doesn't accelerate up to that speed, but instead is born breaking the rules? Physicists have named such defiant speeders tachyons. They remain hypothetical, but they keep popping up in a variety of theoretical contexts, such as in string theory, and are a sign that a mathematical solution isn't stable. Tachyons were boosted into the limelight in September 2011 when physicists were surprised by a result from an experiment under a mountain in Grand Sasso, central Italy, that indicated that a particle called a neutrino had outpaced light, reaching a speed of 299,798,454 m/sec (983,590,728 ft/sec). Tachyons were briefly wheeled out as a possible explanation, but then the team reported flaws in their equipment: a fibre optic cable attached wrongly, a clock ticking too fast. Panic over: nothing had exceeded the speed of light after all. The reason for the disquiet was that the existence of tachyons poses a number of problems, and would mean rewriting the textbooks. For example, if a particle could travel faster than light, it would also be able to move backwards in time and violate causality – the key principle of science that states cause must happen before effect. Despite the problems, physicists have looked for tachyons to no avail, so if they do exist they must interact with ordinary matter so weakly that we can't detect them. So they wouldn't be much use for faster-than-light communication, or time travel – or anything much. Apart from having a cool name. Tachyon.

Sophie Hebden

What Then?

String theory predicts that tachyons have imaginary mass, a bizarre outcome of the mathematics. However, more recent ideas allow tachyons to have real mass, and exist as a field pervading the universe. They may have had a key role in the early universe, perhaps driving the accelerating inflation of space-time that astronomers have observed but don't yet understand: dark energy. Such speculative ideas within string theory – in which certain energy conditions of the tachyon field result in dark energy – are called tachyon cosmology.

What Gives?

'Speedy' Literal translation of Greek word tachus, after which the tachyon is named.

1960s Era in which the theoretical framework for tachyons was developed.

Antitachyon Antiparticle counterpart of the tachyon. Tachyons can be assigned the normal properties of matter, such as charge or spin.

What Else?

What if time ran backwards? *See page 42*

What if everything were made of string? *See page 84*

C

WHAT IF YOU HIT WARP SPEED?

Brian Clegg

The universe is a big place – if we were ever to explore it, we would need to travel fast enough to cross vast distances. And that means travelling faster than light. If you are limited to light speed, for example, it would take 2.5 million years to reach our nearest neighbouring galaxy in Andromeda. However, this presents a serious problem, as special relativity seems to put an insuperable barrier in place. Nothing can move through space faster than the speed of light. As a body gets close to light speed, its mass increases to the extent that on reaching the velocity of light the mass would become infinite. Luckily, there is a potential get-out clause. If you manipulate spacetime itself, there is no problem with the light-speed limit. In its inflationary phase, for example, the universe is thought to have expanded vastly quicker than the speed of light. And quantum particles regularly tunnel from one place to another taking no time to cross the intervening space, effectively moving infinitely quickly. Science fiction has long presented the answer to this conundrum in the form of a warp drive, a spaceship engine that distorts spacetime itself, making it possible to sneak around the light-speed barrier. This was long thought to be nothing more than fiction, but in 1994 Mexican theoretical physicist Miguel Alcubierre wrote a paper entitled 'The Warp Drive: Hyper-Fast Travel Within General Relativity'. It had long been realized that making a tear in spacetime – a so-called wormhole – would in principle make it possible to exceed light speed, but Alcubierre's design worked by expanding spacetime behind the ship and contracting it in front, producing a self-contained warp that could, in principle, propel the ship far faster than the speed of light. The only catch was that like wormholes, the drive needed a hypothetical substance called exotic matter that effectively had negative mass to be stable – in quantities comparable in scale to the planet Jupiter.

What Then?

In 2011 American NASA scientist Harold White found a way to modify the shape of the warp bubble that would reduce the requirements for negative mass to around 2 tonnes. If it is possible to produce such exotic material, this seems more achievable. There are tiny effects such as the Casimir effect that arises when two metal plates are very close together that have the same effect as negative mass, but as yet this is still a major problem to crack before we can engage warp.

What Gives?

0.003694 Percentage of the speed of light travelled by the fastest ever human beings when the crew of *Apollo 10* flew at 39,897 km/h (24,791 mph) relative to the Earth.

114,776 Years it would take to fly to the nearest star to the Sun (Proxima Centauri) at the speed of *Apollo 10*.

What Else?

What if gravity were not a force? *See page 48*

What if there were extra dimensions? *See page 52*

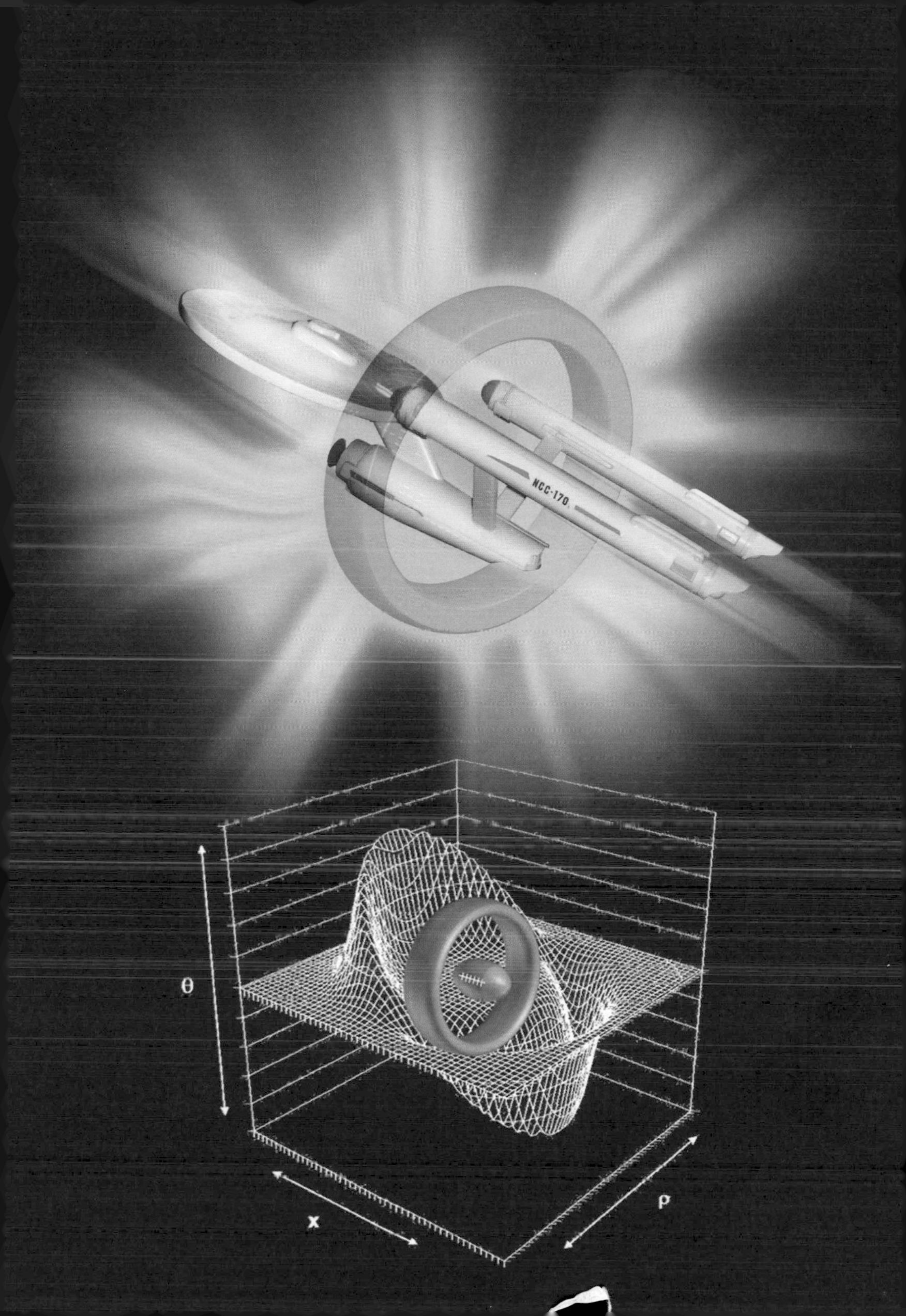
NCC-170
θ
x
ρ

WHAT IF GRAVITY WERE NOT A FORCE?

Rhodri Evans

We tend to think of gravity as a force, like the force of attraction between a positively and a negatively charged particle. But in fact it seems to be something much stranger – a warping of space and time. In 1907 Einstein had what he would later refer to as the 'happiest thought of my life'. Two years after publishing his special theory of relativity, Einstein started thinking about how acceleration would change the ideas he had developed in that theory. In a flash of insight, he realized that there was no way to tell the difference between acceleration and gravity, and over the next nine years he developed his general theory of relativity, one of the greatest scientific achievements ever. The general theory of relativity leads us to think of gravity in an entirely different way. Rather than a force of attraction between two bodies, gravity – in Einstein's radical theory – is actually a property of spacetime itself. Bodies with mass distort the very fabric of spacetime: the greater the mass of the object, the larger this distortion is. Einstein predicted that light itself would be deflected by gravity, a prediction that was shown to be correct by English astrophysicist Arthur Eddington in 1919 when he measured the positions of background stars during a solar eclipse. This effect is used every day in so-called 'gravitational lenses', which enable us to view distant and faint galaxies whose light is lensed by the gravitational deflection and amplification of foreground objects. The distortion of spacetime produced by masses leads to time passing more slowly where spacetime is more distorted. This effect has to be allowed for in GPS systems, since time passes more quickly where the satellites are located than it does for us on the Earth's surface (see page 38). At the event horizon of a black hole, the boundary beyond which light cannot escape, time actually stands still. It seems that gravity is woven into the very fabric of the universe.

What Then?

One remarkable implication of general relativity is that it may be possible to create wormholes, which connect different parts of the universe. If we could survive a trip through a wormhole, we could travel virtually instantaneously to an entirely different part of the universe.

What Gives?

420 million years

Age after the big bang of the most distant galaxy yet seen in the universe. NASA announced the detection of the galaxy, 13.3 billion light years away, in November 2012. It could only be seen because it was gravitationally lensed by a foreground cluster of galaxies, amplifying its incredibly faint light.

2.9 km Distance of boundary or event horizon from the centre of a black hole of solar mass 1 (a unit for measuring the mass of stars, equivalent to the mass of our Sun). Any object can be a black hole, provided it is sufficiently dense. If our Sun were replaced by a black hole of the same mass, the Earth's orbit would be unaffected. (2.9 km is 1.8 miles.)

What Else?

What if Einstein was wrong? *See page 78*

What if space and time came in loops? *See page 86*

HISTORICAL
WHAT IF YOU CAN'T TELL IF YOU'RE MOVING?

Mention relativity and the name that instantly springs to mind is Albert Einstein. But the groundwork on relativity was performed 300 years earlier by the great Italian physicist and astronomer Galileo Galilei. This has nothing to do with Galileo's astronomy and his support for the Copernican theory that the Earth is not at the centre of the universe, which led to his infamous trial on suspicion of heresy. Instead it was part of his magnificent contribution to physics.

Galileo put forward the simple but revolutionary idea that if we are in steady motion, we have no way of telling that we are moving, without looking outside. He envisaged setting up a physics laboratory on a ship that was making steady progress on a smooth sea. In the enclosed laboratory, with no windows, it would be impossible to determine whether or not the ship was moving. As far as the experiments were concerned, everything was still.

At the heart of Galileo's relativity was the idea that movement is not absolute – we always have to be aware what we are moving with respect to. Let's assume that you are sitting in a chair as you read this book. You aren't moving with respect to your seat. In many cases you won't be moving with respect to the Earth – though this would not be true if you are in a car, or on a train, plane or boat. However, even fixed in position on the Earth, you are spinning around with the planet's rotation, shooting around the Sun on the Earth's orbit and hurtling across space with the Milky Way galaxy towards our nearest intergalactic neighbour, the Andromeda galaxy. The concept of motion is inherently relative.

Galileo is said to have demonstrated this dramatically to a group of his friends. He took them out on Lake Piediluco in Umbria, in a fast boat rowed by six men. When they had got up a good speed, Galileo asked if anyone had anything heavy on them. One friend, Stelluti, produced the door key of his home, a massive piece of complex shaped iron, the only key to his handmade lock.

Taking the key, Galileo hurled it straight up into the air. Stelluti was horrified. It seemed obvious that the fast-moving boat would slip away from under the key before it fell back down. He was going to lose his only key in the dark waters of the lake. Galileo's other friends managed to restrain Stelluti from hurling himself off the back of the boat into the lake in an attempt to catch the key. The key fell back down into Galileo's lap. As Galileo knew, the fact that the boat was moving at high speed over the water was irrelevant. From the key's viewpoint, the boat was not moving. If the key was thrown straight up it would fall back down at the same point in the boat.

The concept of relativity was deeply shocking to Galileo's contemporaries. It seemed unnatural. Yet it would have been impossible to produce any of the basic physics of motion that followed without it. English physicist Isaac Newton owed a huge debt to Galileo (as Einstein would also in his time). If two cars head towards each other, each going at 80 km/h (50 mph) relative to the ground, the result will be a 160-km/h (100-mph) head-on crash. Turn a car round so that both travel in the same direction and as far as one car is concerned, the other isn't moving. You could step from one to the other without harm if they were close enough. Relativity is the key to understanding motion.

WHAT IF THERE WERE EXTRA DIMENSIONS?

Frank Close

If extra dimensions exist, it might be possible to take shortcuts through space and time and find the theory of everything. Some particle physicists believe there could be more dimensions than we are normally aware of; the 'higher' dimensions could be extremely small, unlike those of space and time, which have grown to encompass the visible universe for 13.8 billion years. Superstring theory, which unites gravity and other forces, posits that particles such as electrons and quarks are composed of strings, which exist in higher dimensions. To see these features directly would require experiments far beyond the reach of the Large Hadron Collider (LHC) in Geneva, Switzerland – or any realistic future particle accelerator. However, according to some theories one of these dimensions may be within the reach of the LHC and, if so, its presence might be revealed. To imagine what might occur, suppose we were creatures living on a table top, only aware of two dimensions: if something fell off the table, it would seem to our senses to have spontaneously disappeared; if a feather floated down onto the table, it would appear as from nowhere. The spontaneous appearance or vanishing of certain particles at the LHC could be one way that higher dimensions are revealed. Even though the dimensions are small, they can be accessed at any point in our known universe, so their total effects could be huge. The electromagnetic, weak and strong forces are, with gravity, the fundamental physical forces: if the electromagnetic, weak and strong forces act only in our conventional dimensions while gravity can 'leak' into all of them, this could explain several mysteries. For example, it could explain why gravity appears so feeble to us: we only feel the 'remnants'. Some theories of dark matter suggest that while stuff might be trapped in higher dimensions, its gravity leaks into our dimensions, so that we can feel its presence but not see it. At present these ideas are awaiting experimental proof.

What Then?

If space is curved around in a higher dimension, we could take shortcuts via that dimension. The proposed curving of space is analogous to folding a sheet of paper so that its opposite edges touch one another: such an arrangement would enable 'instant' travel 'across' from the top half to the bottom half of the folded over sheet. In a world such as this, some exploits of science fiction could become science fact.

What Gives?

10^{-35} m The extent of the higher dimensions proposed by superstring theory.

10^{19} GeV Energy of particle collisions that would be required to detect the tiny distances described above. This is around 1 million billion orders of magnitude larger than can be achieved at the LHC.

What Else?

What if everything were made of string? *See page 84*

What if there were no dark matter? *See page 112*

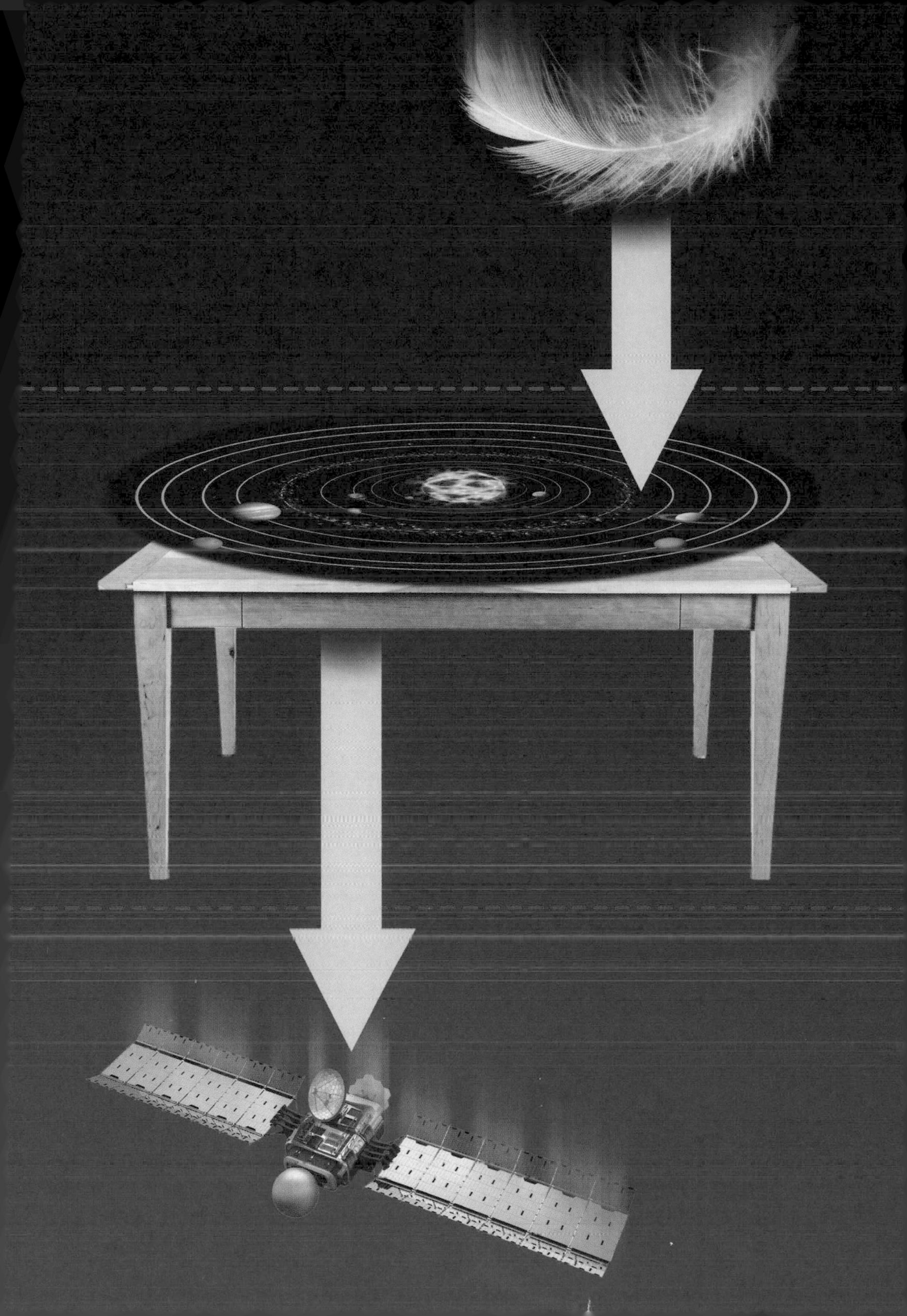

PARTIC
PHYSIC

PARTICLE PHYSICS

PARTICLE PHYSICS
INTRODUCTION

The ancient Greeks had two opposing theories of matter. The prevailing idea that held sway through to the beginnings of modern science was Empedocles' concept that everything was composed of four elements – earth, air, fire and water. But there was an alternative theory that you could cut matter up into smaller and smaller pieces until it was 'uncuttable' or a-tomos. Supporters thought that everything was made of atoms.

This idea was revived by English scientist John Dalton when he devised modern atomic theory in the early 1800s. To begin with most scientists thought that atoms were just a convenient way of explaining the chemistry. It wasn't until Einstein's day that atoms began to be taken seriously as real entities, forming an early precursor of particle physics.

Another contribution came from the time of the great English physicist Isaac Newton in the 17th-18th centuries, by way of his attempt to understand light and gravity. Newton thought that light was made up of tiny particles called corpuscles, and though he did not formally try to explain how gravity worked, he seems, like many of his contemporaries, to have assumed that it involved bodies being bombarded by a flow of particles that made the apparently ethereal force act at a distance.

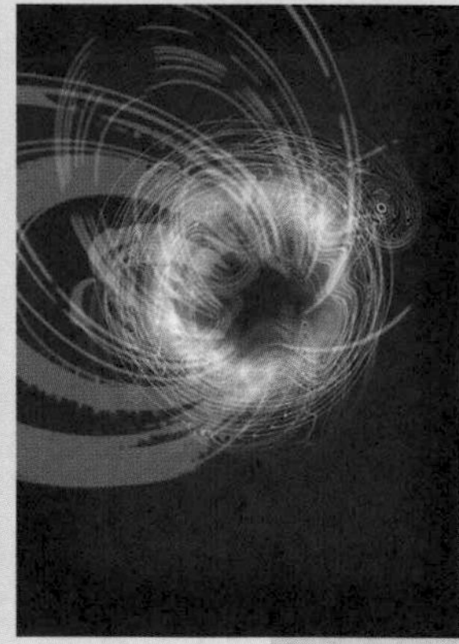

Newton's ideas on particles proved wrong in detail, but they give a feel for the basis of particle physics that everything – matter and forces alike – is the result of the interaction of tiny particles. Our understanding of the actual particles involved has changed over time. At the start of the 20th century it was thought that the fundamental particles of matter were atoms. These were discovered to be made up of electrons, protons and neutrons. More recently protons and neutrons themselves were found to have constituents called quarks. As yet quarks and electrons seem truly fundamental.

Now we add in a collection of particles found as a result of nuclear interactions and cosmic rays – neutrinos and muons, for instance – plus the particles that carry the forces, like photons for electromagnetism, and the joker in the pack, the Higgs boson. Together, we have a particle zoo called the standard model, our best current picture of the fundamental components of everything.

Particle physicists are sometimes likened to children, attempting to understand a clock by hitting it with a hammer and measuring what components are produced. Huge accelerators like the Large Hadron Collider at CERN in Geneva, Switzerland, certainly provide the sledgehammers of physics – but they do help reveal the secrets at the heart of reality.

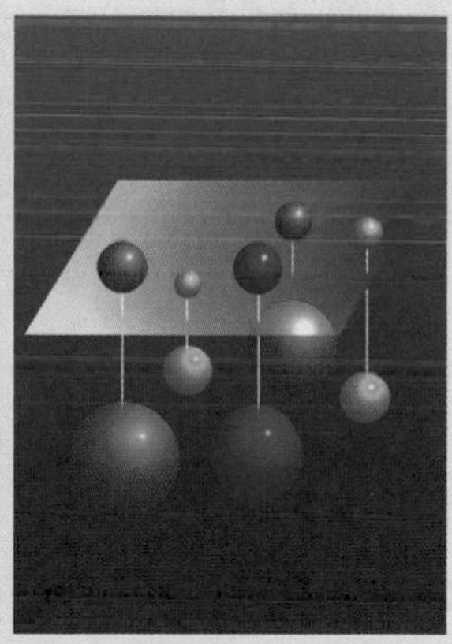

WHAT IF WE HAD GUTS AND TOES?

Frank Close

Physicists hoping to bring gravity into line with the other forces of nature are seeking a Theory of Everything (TOE), while others attempt a unification of the strong forces that hold together the nuclear particles of an atom and the already united electromagnetism and weak force – the Grand Unified Theory (GUT). With a GUT in place, you would understand why your hair doesn't go up and down as you breathe. If your hair was full of static electricity, the repulsion of like charges would push the individual hairs apart – sparks might also fly! This doesn't happen because matter can be electrically neutral, allowing your hair to stay in your chosen style. But this self-evident fact is quite remarkable. Inside our atoms are many electrically charged particles: electrons in the outer reaches of the atoms, protons densely packed in the central nucleus and powerful electrical fields fill the entire atomic volume. Protons and electrons are very different – protons are nearly 2,000 times more massive than an electron and far more spread out in space. Yet the positive electric charge of a proton balances the negative charge of the electron so precisely that matter is able to be electrically neutral, enabling gravity to dominate at large distances. This counterbalancing is so precise that it is believed to be fundamental, not mere accident. The mystery deepens when we realize that the proton is made of smaller particles: quarks. Each quark carries electric charges that are either $+2/3$ or $-1/3$ fractions of a proton's charge. And quarks cluster in groups of three, not five or even alone; the conspiracy of three times $1/3$ precisely counterbalancing the negative unit of electric charge on the electron makes the electrical mystery even more intriguing. The electron, with electric charge -1, knows nothing of quarks – it is, as far as we can tell, fundamental, and not made of yet smaller particles. So the neutrality of matter suggests there is a unification between the strong and electromagnetic forces, and between quarks and the electron.

What Then?

If physicists found a theory of everything, it might actually not solve very much in practice. We probably already have the theory of everything for chemists and biologists in Dirac's equation, which describes how the electron behaves in and around atoms. However, in practice it's only possible to solve the theory in a few cases, such as for simple atoms like hydrogen. Having the equations on a T-shirt is one thing; solving them is another.

What Gives?

1964 Year in which the theory that protons and neutrons are made of quarks was first proposed, by American physicist Murray Gell-Mann and his Russian-born colleague George Zweig.

1975 Year in which the first version of a Grand Unified Theory was proposed, by American physicists Howard Georgi and Sheldon Glashow.

What Else?

What if everything were made of string? *See page 84*

What if there were no dark matter? *See page 112*

WHAT IF THERE WERE SUSY?

Frank Close

Every fundamental particle comes in one of two varieties, fermion or boson. The two classes behave very differently. Fermions are like cuckoos, where two in the same nest is one too many, while bosons are like penguins: the more the merrier. Photons of light are bosons, and intense laser beams are an example of large numbers of bosons acting collectively. Electrons are examples of fermions, and it is their mutual incompatibility that gives rise to the structure of atoms and matter. The atoms of heavy elements contain many electrons, but each one has to find a unique state, in accordance with the laws of quantum mechanics. So two atoms cannot exist in the same place, which is a prerequisite for structure. One unproven theory of particle physics, supersymmetry or SUSY, suggests that every known fermion would have a boson sibling at the same mass, and likewise, every known boson would have a fermion counterpart. So the familiar electron, with negative charge, would have a SUSY counterpart – a boson with negative charge, known as a selectron – and with the same light mass as an electron. The positron would likewise have a positively charged SUSY partner: the supositron. Bosons with opposite electric charge attract one another. With no restriction on the numbers of bosons in the same state, lightweight selectrons and supositrons would clump in balls of electrical energy before annihilating in a burst of photons, destroying the quiescent environment within which electrons form atoms. As this doesn't happen, if SUSY exists the symmetry must be broken, implying that the theory does not apply completely to the real world. If SUSY is real at all, it seems that the super-particles must be very massive compared to those that make our familiar stuff. Experiments at the Large Hadron Collider (LHC) at CERN in Geneva, Switzerland, are looking for heavy SUSY particles. No examples of SUSY particles have yet been seen.

What Then?

Known particles explain the structure of at most 5 per cent of the material universe. The rest is made of something else, including large clumps of 'dark matter', which is sensed only by its gravitational pull on the visible galaxies. SUSY theories propose the possibility that there are massive, electrically neutral 'dark particles', which might be stable enough to seed the dark matter.

What Gives?

10^{19} Number of times more massive than a proton that a proposed SUSY particle could be, if superstring theory is correct.

2010 Year in which the LHC at CERN began looking for SUSY particles.

Half Proportion of particles scientists have already discovered, if SUSY theories are correct.

What Else?

What if you made a quantum leap? *See page 16*

What if the Higgs boson did not exist? *See page 66*

WHAT IF
ANTIMATTER FELT ANTIGRAVITY?

Frank Close

If antimatter felt antigravity, it might explain why the material universe exists. One of cosmology's biggest puzzles is why the universe is made of matter to the exclusion of antimatter. Theory predicts, and experiment confirms, that every variety of fundamental particle has an antimatter analogue, with the same mass but the opposite electric charge. The antiparticle of the negatively charged electron is the positively charged positron; the antiproton has a negative charge. The most dramatic property of antimatter is that when an antiparticle meets its particle analogue they can mutually annihilate in a flash of radiation, such as photons of light. In reverse, high-energy radiation can convert into counterbalanced numbers of particles and antiparticles. This has been seen in experiment, and leads to the theory in which the huge energy in the immediate aftermath of the big bang converted into particles of matter and antimatter in perfect symmetry. Yet the observable universe consists of matter; there is no evidence for large-scale clusters of antimatter. How did it all disappear? Is it that matter and antimatter are not perfect mirror images: perhaps most matter and antimatter annihilated each other, but a small excess of matter led to the current material universe. However, no evidence for this has been found. If matter and antimatter repel one another by gravity, it may be that we live in a 'multiverse', and a vast universe of antimatter was repelled by newborn bulk matter post-big bang; we occupy the matter part of this multiverse, and its antimatter counterpart is far away, undetected. Gravity acts so weakly on individual particles of matter or antimatter that its effects are truly difficult to measure. Individual easily available antiparticles, such as positrons, have electric charge, and stray electric and magnetic forces in surrounding matter swamp gravity's feeble effects. Even if antimatter seems to feel the same gravitational attraction as matter, the uncertainties are huge: the possibility that antimatter feels antigravity is still open.

What Then?

The AEgIS experiment at CERN in Geneva, Switzerland, plans to become the first to measure the direct effect of the Earth's gravitational force on antimatter. (AEgIS stands for 'Antihydrogen Experiment: Gravity, Interferometry, Spectroscopy'.) Atoms of antihydrogen are electrically neutral so the effects of gravity on them may be measurable. In the experiment, beams of these atoms will pass through narrow gratings and researchers will measure how far the antihydrogen atoms have fallen during their flight.

What Gives?

$9.10938291(40)\times10^{-31}$ kg
Mass of a positron, making it difficult to detect the tiny effect of gravity.

10^{40} The number of times weaker the gravitational force in a hydrogen atom is to electrostatic, meaning electromagnetic effects swamp gravitational on the scale of a particle of antimatter.

What Else?

What if the big bang theory were wrong? *See page 82*

What if antimatter were everywhere? *See page 92*

HISTORICAL
WHAT IF ATOMS HAVE COMPONENT PARTS?

The theory that everything is made of atoms goes all the way back to ancient Greece. One school of ancient Greek thought held that you could cut matter into ever smaller pieces until you reached a stage at which it was a-tomos ('uncuttable') – from which we get the name 'atom'. But it wasn't until English scientist John Dalton discovered atomic weight at the beginning of the 19th century that the idea of the atom began to have scientific relevance. Dalton's breakthrough led to an understanding that each element is composed of its own atoms, which in turn give the element its chemical distinctness and characteristic weight. Nevertheless the existence of the atom was still considered conjectural.

When English physicist J. J. Thomson announced the discovery of the electron in 1897 the break from the ancient Greek concept was complete. In 1904, he proposed the 'plum pudding' model of the atom. In this model, negatively charged electrons (plums) dotted throughout the atom were surrounded by a sort of positively charged cloud (the rest of the pudding), thereby producing an overall neutral charge. Thomson initially felt there must be nearly 2,000 electrons in a hydrogen atom because he felt they were what gave it its mass.Two years later, Thomson radically downsized the number of electrons to a number roughly equal to an element's atomic number.

Like a real pudding, the model didn't last long. In 1909, Thomson's former student Ernest Rutherford performed a now celebrated experiment in which he fired alpha particles at a strip of gold foil – and a very small number of the particles bounced back. (Alpha particles consist of two protons and two neutrons, and are identical to a helium nucleus, although Rutherford did not know what they were at the time since protons and neutrons had not been discovered.) Rutherford announced a new model of the atom in which a tiny positively charged nucleus, contained within a fraction of the volume of the overall atom, was responsible for almost all of the atom's mass – while the electrons orbited it. It was possible to understand that in Rutherford's experiment the majority of alpha particles would pass through the atoms of the gold foil but that those that hit the nuclei would bounce back.

In 1920, Rutherford suggested the name proton for the positively charged particles in an atom's nucleus, and conjectured that there also existed a neutral particle, subsequently named the neutron. So began a trend – as particle physics progressed over the following years, theory now preceded discovery. It wasn't until 1932 that English physicist James Chadwick proved the existence of neutrons.

We now have the structure of the atom that we learn about at school – a nucleus, containing protons and neutrons, orbited by electrons. But this is just the first exhibit in what has been appropriately called a 'particle zoo'. In the early 1970s, scientists came together to develop what has become known as the Standard Model of particles and forces. It proposes that everything in the universe is made from 12 fundamental particles and four fundamental forces.

The model consists of three groups: quarks, leptons and gauge bosons, as well as the Higgs boson. Quarks come in six 'flavours': up (u), down (d), charm (c), strange (s), top (t) and bottom (b). Protons consist of u, u and d quarks. Neutrons are u, d and d quarks. Electrons are examples of leptons, of which there are also six flavours: electron (e–), electron neutrino (ve), muon (u–), muon neutrino (vu), tauon (t–) and tauon neutrino (vt). Gauge bosons carry three of the four fundamental physical forces (the weak, the strong and the electromagnetic forces); the fourth fundamental force is gravity, which isn't explained by particle physics.

One can understand why Italian physicist Enrico Fermi felt compelled to utter 'young man, if I could remember the names of these particles, I would have been a botanist'.

WHAT IF THE HIGGS BOSON DID NOT EXIST?

Frank Close

The short answer is: we wouldn't exist. The Higgs field, which we believe permeates the universe, gives mass to fundamental particles – such as the electron, quarks and the W boson. The size of an atom is governed in part by the mass of the electron; if it were lighter, atoms would be larger than in reality. If the electron had no mass then atoms would be infinitely big, which means that atoms would not exist. So there would be no chemistry, biology or life. The mass of quarks causes the strong nuclear force (which holds together the particles of the nucleus) to act only over a very short range. This is why atomic nuclei are very compact. If the quarks had no mass, atomic nuclei would not form. However, individual neutrons and protons would exist, and have masses similar to what occurs in reality. Contrary to a popular misconception, the Higgs does not give mass to everything in the universe; in fact, it is responsible for only a small part of the whole. Most of your mass, and that in everything we see, is due to the protons and neutrons in atomic nuclei, and their masses come from the kinetic energy of the quarks entrapped within these particles. This would exist even without any Higgs. The W boson is the agent that mediates the weak force, which is involved in radioactive decay and converts hydrogen into helium in the heart of the Sun. The large mass of the W is what makes the Sun burn very slowly. Evolution has taken billions of years; but for the Higgs field, the W would have had no mass, the weak force would have been powerful and the Sun would have burned out long ago – assuming that it even existed in the first place, which it probably wouldn't have! Without the Higgs, there would be a universe, but not as we know it.

What Then?

The idea that fundamental particles get their mass this way was proposed by several people, independently, around 1964. So the convention of referring to the 'Higgs field' is unfair to these others. However, the Higgs boson is fairly named, because English theoretical physicist Peter Higgs alone drew attention to a consequence of the theory: there should exist this unstable massive particle, which can be used to test the theory in experiments. The particle was isolated in 2012.

What Gives?

125 GeV Mass of Higgs boson – about 130 times heavier than a hydrogen atom. No particle physics accelerator was capable of producing this massive particle until the 21st century. (The giga electron volt is a unit used by particle physicists: 1 GeV is the equivalent of 1.783×10^{-27} kg.)

2012 Year in which the Higgs boson was found by experiments at CERN, Geneva, and also at Fermilab, near Chicago.

What Else?

What if there were SUSY? *See page 60*

What if everything were made of string? *See page 84*

WHAT IF WE COULD SEE THE ATOM?

Simon Flynn

An atom is 1 billion times smaller than an orange, the same difference in size as that between an orange and Jupiter. The smallest resolution our eye can see is a fraction less than the width of a single strand of hair. This is still about 1 million times bigger than an atom. Our ability to see the world in greater detail took a leap forward in the 17th century when Dutch scientist-tradesman Antonie van Leeuwenhoek invented the optical microscope. Over the years, while microscopes improved dramatically they never came close to allowing us to see an atom because the wavelength of visible light is far bigger than an atom. You can't detect an object where two points on it are less than a wavelength apart as the light reflecting from the points interferes with each other. However, researchers at the University of Manchester, England, demonstrated in 2011 a new type of optical microscope that permits them to see individual molecules, thanks to the use of glass beads just a few nanometres wide that allow the wavelength limit to be overcome. But as yet we are still unable to see atoms directly. During the 1980s, a technique for seeing individual atoms indirectly was developed by German physicist Gerd Binnig and his Swiss colleague Heinrich Rohrer, both at IBM; this won them the Nobel Prize in Physics in 1986. Scanning tunnelling microscopy (STM), as it became known, works by bringing a fine tungsten needle (its tip just one atom) to within a nanometre of the surface of the object being investigated then moving the needle across the object's surface. The very small gap between needle and object and a feature of quantum mechanics known as quantum tunnelling enable electrons to jump from needle to object, causing a small current to flow between the two. As the needle moves across the surface, the tip is moved up or down to keep the current constant. From this, an atomic picture of the object's surface can be built up, with the position and size of each individual atom detailed.

What Then?

In 1990, scanning tunnelling microscopy (STM) was shown to be capable of moving and placing individual atoms. Scientists at IBM used STM to drag and drop 35 individual atoms of xenon distributed randomly on a nickel surface so that they spelled out the company's name. The prestigious scientific journal *Nature* featured the resulting image on its front cover.

What Gives?

400–700 Wavelength of visible light in nanometres.

1981 Year in which Gerd Binnig and Heinrich Rohrer developed scanning tunnelling microscopy (STM).

275 Factor of magnification achieved by Dutch scientist Antonie van Leeuwenhoek's most powerful surviving microscope.

What Else?

What if atoms had component parts? *See page 64*

What if something were colder than absolute zero? *See page 118*

WHAT IF WE COULD REPLACE ELECTRONS WITH LIGHT?

Brian Clegg

American businessman Gordon Moore, co-founder of the Intel computer chip manufacturer, observed in 1965 that the number of components in integrated circuits had doubled in every recent year, and he predicted this would continue for another 10 years or more. With a subtle modification to doubling every two years, 'Moore's law' has proved a remarkably good fit for more than four decades. But physics makes it clear that this increase can't continue forever. There comes a point where the connections within the chip are down to the order of the size of atoms, leaving no room for miniaturization. And since the electrons that flow around the chip are quantum particles, they are increasingly likely to be able to tunnel from connection to connection as the scale becomes smaller. One obvious way to escape the restriction is to move from two dimensions to three. This can be done to a degree, but the complexities of arranging connections in three dimensions rapidly ends up with further limitations. Yet it is possible that future generations of computer could involve changing the quantum particles at the heart of the machine from electrons to photons of light – moving from electronics to photonics. Electrons and photons are very different types of particle. Electrons have mass, photons don't. Electrons have an electric charge, photons don't. However, as quantum particles they both can have wavelike properties and both can be used for the essential job of carrying a signal. Building a light-based computer presents many practical challenges, but provides one huge advantage. Electrons repel each other and are fermions – particles that obey the Pauli exclusion principle, which states that no two can be in exactly the same state. Photons are bosons, particles that love to congregate. They can pass through each other and as many as you like can occupy the same space, making it possible to envisage a light-based chip with billions of connections that simply pass through each other.

What Then?

There are ways to produce a light-based transistor, where the transmission of light in one direction is controlled by the intensity of light hitting a special crystal in another direction, but as yet these are relatively large-scale devices. However, there are many special optical devices that only operate on light at the nanoscale, suggesting that with a very different approach a computer in which billions of signals pass through the same point at any time could eventually be feasible.

What Gives?

1953 Year in which the first maser – a microwave precursor to the lasers that would be essential for optical computers – was produced by American physicist Charles Townes.

2.5 billion Number of transistors in the largest computer processor commercially available in 2012, Intel's 10 Core Xeon Westmere.

What Else?

What if we could compute with quanta? *See page 30*

What if robots were conscious? *See page 138*

WHAT IF ATOMS WERE NOT EMPTY?

Frank Close

If atoms were filled with electrons, free to travel where they please, the atoms would collapse in a burst of radiation almost as soon as they formed. An atom consists of negatively charged electrons that encircle a compact, central, massive and positively charged nucleus. It is the rule of electric forces ('like charges attract') that holds the system together. A popular analogy for the structure of the atom is that it is like a miniature solar system, with planetary electrons whirling around a central nuclear sun. This is misleading for many reasons – not least that it does not do justice to the atom's emptiness. The Earth's orbit is roughly 100 times the diameter of the Sun, which is in turn about 100 times that of the Earth itself; the analogous figures for the size of a hydrogen atom, the central nucleus and an electron are nearer to 10,000, making atoms remarkably empty, even compared to the solar system. In addition, the strength of electric and magnetic forces within atoms are much stronger than those of gravity, and the atom is much, much smaller than the solar system. Conventional physics suggests that an electron should spiral into the nucleus within a fraction of a second, in a burst of light – that this does not happen shows that the laws that operate within atoms (those of quantum mechanics) are not those of which we are normally aware. Instead of being able to go where they please, electrons are restricted – like someone on a ladder who can only step on individual rungs. Each rung corresponds to a state in which an electron has a given amount of energy. When an electron moves down the ladder from a rung with high energy to one with lower energy, the difference in energy is radiated as a photon of light. The characteristic bar code of colours – the spectrum of this radiation – is unique to each particular element.

What Then?

The spectrum of light from distant objects in the universe can reveal which elements are present. Electrons are fermions (see page 58) and so two electrons cannot be in the same state at the same time. Thus individual atoms maintain their identities, which gives matter its structure. Neighbouring atoms may exchange electrons, or one may borrow an electron from another; the result is to grip the two atoms to one another, creating molecules.

What Gives?

1897 Year in which the electron was discovered by English physicist J.J. Thomson.

1911 Year in which New Zealand-born physicist Ernest Rutherford published his theory of the atomic nucleus.

1913 Year in which Danish physicist Niels Bohr completed the model of the atom.

What Else?

What if you made a quantum leap? *See page 16*

What if we could see the atom? *See page 68*

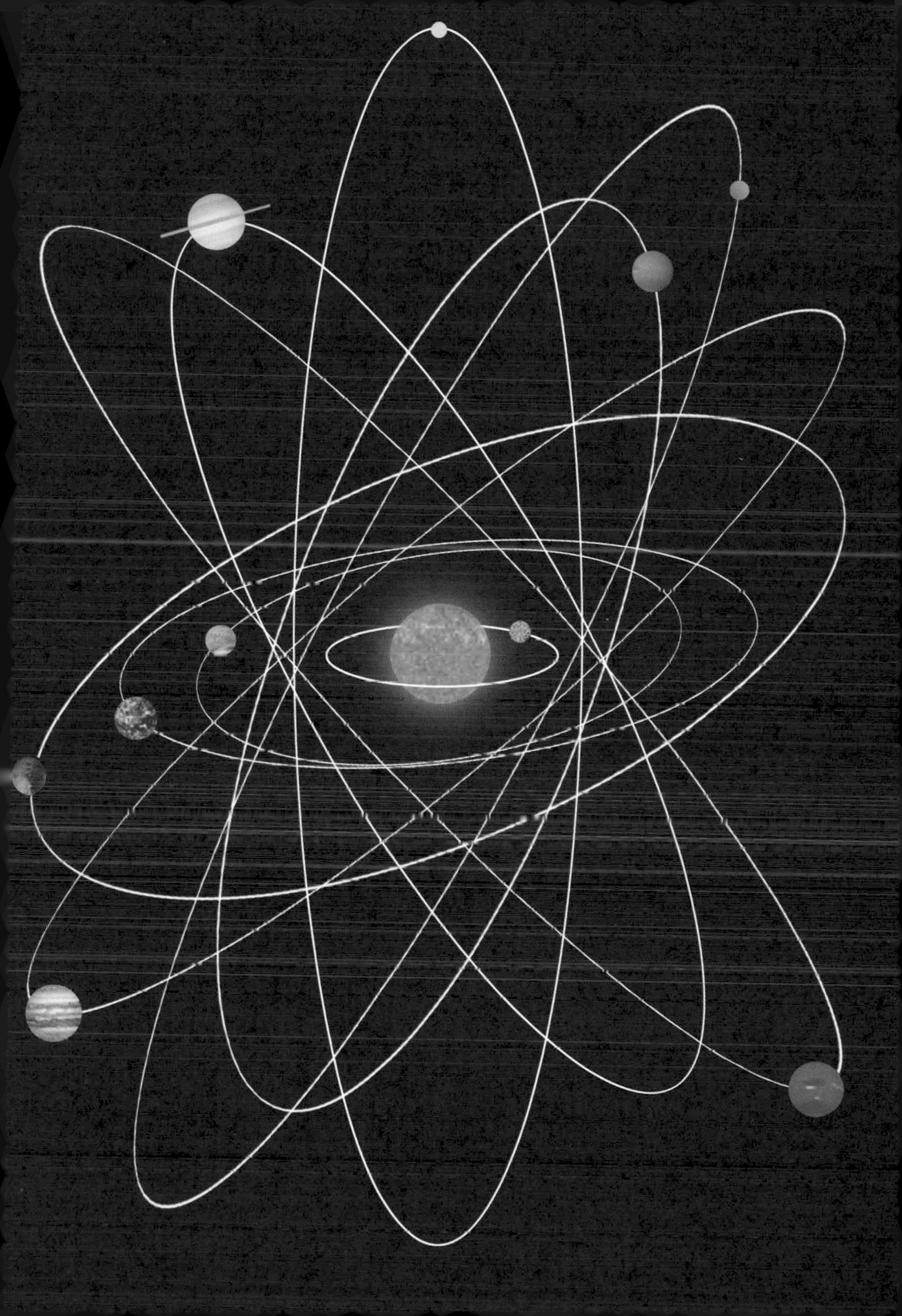

COSMO

COSMOLOGY

COSMOLOGY INTRODUCTION

There is an old saying: 'There's speculation, then there's wild speculation, and then there's cosmology.' This may be unfair, but it reflects a truth. Cosmology is about the origins and nature of the universe as a whole – and this is inherently difficult to study and because of the scale of the universe, making anything but our nearest neighbours impossible to visit, this is inherently difficult to study.

Most physicists can undertake an experiment in a laboratory, even if some have to resort to such magnificent 'laboratories' as the Large Hadron Collider at CERN (the European Organization for Nuclear Research in Geneva, Switzerland). But we can't experiment with the universe. Another mainstay of science is repeatability. But the universe we experience is a one-off. You can't set it going again to compare results.

All the cosmologist can do is rely on observations, and those observations are often very indirect. It isn't possible to get our hands on a galaxy, for example – all that we know is based on information carried to us by light. Until recently scientists relied purely on visible light, but now they can employ the whole electromagnetic spectrum – from radio, through microwaves, infrared, visible, ultraviolet, X-rays and gamma rays. This has opened up a vast extra flow of information, but the observations remain indirect.

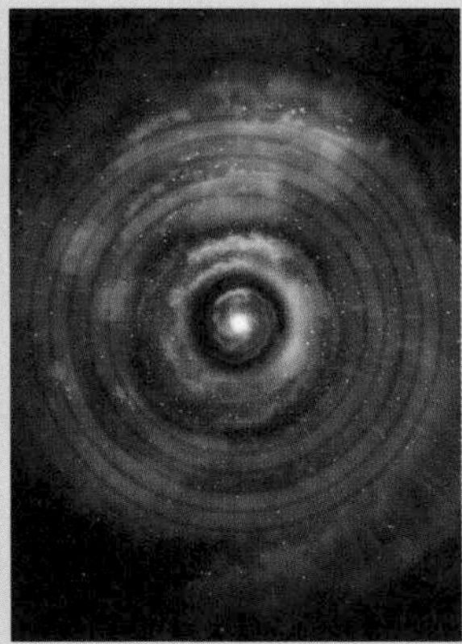

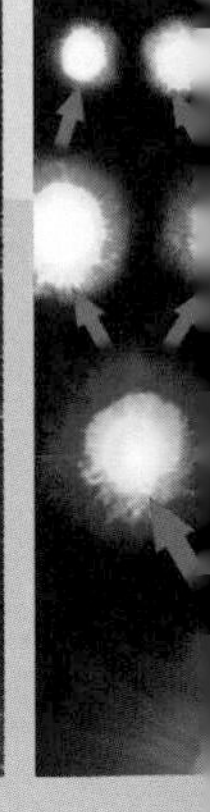

Another problem for scientists is time. In some respects, cosmologists are luckier than Earth-bound scientists who try to understand what happened millions or billions of years ago. A geologist or palaeontologist can only go on whatever remains are discovered in the ground, trying to deduce what time period they come from. Cosmologists, though, have a visual time machine. The further they look out into the universe, the further they look back in time. Because the light that carries information to us travels at a finite speed, our views of space are also views of the history of the universe.

Look, for instance, at the Andromeda galaxy, the most distant object you can see with your naked eye. This is around 2.5 million light years away, meaning we see it as it was 2.5 million years ago. The most distant direct observations take us back over 12 billion years into the past. But it is only indirect observations like the cosmic microwave background that give us insights into the universe in its extreme youth, making any attempts at describing the origins of the universe speculative at best.

More than any area of science, our best cosmological theories could easily be overthrown. There might not even have been a big bang. The universe still has plenty of surprises waiting for us.

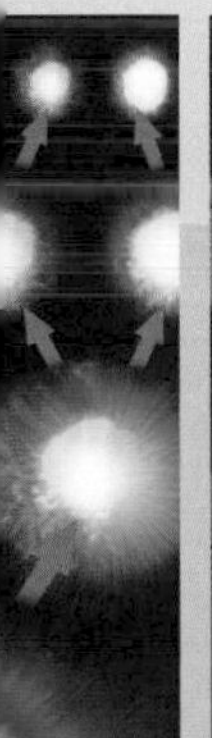

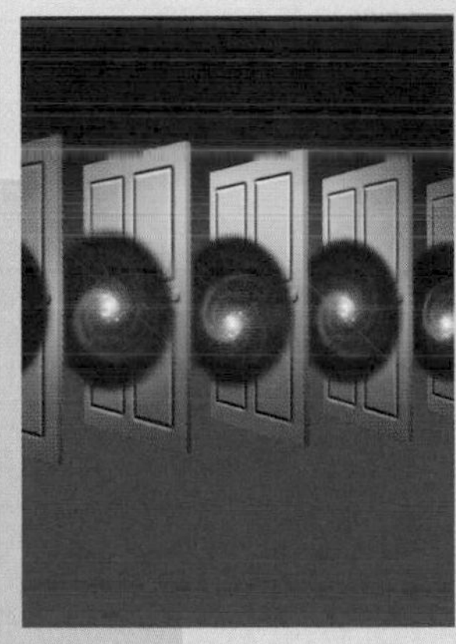

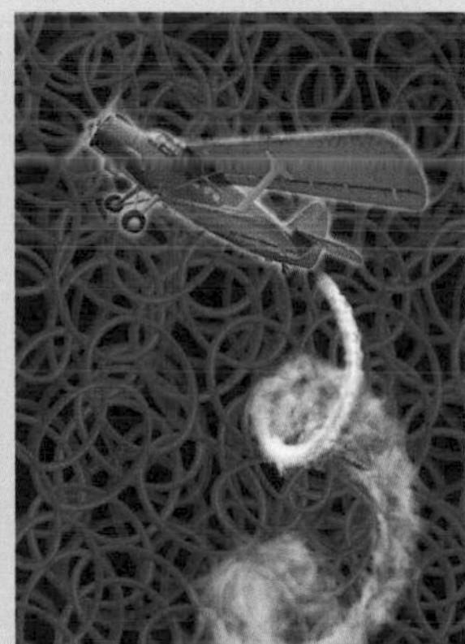

WHAT IF EINSTEIN WAS WRONG?

Rhodri Evans

'My greatest blunder' is how Einstein referred to the cosmological constant he introduced into his equations of gravity. He posited the cosmological constant at a time (in 1916–17) when most astronomers thought the universe was static, neither expanding nor contracting. Its job was to 'fix' his equations to fix the size of the universe. But in 1929 American astronomer Edwin Hubble discovered that the universe was expanding. This led to Einstein's famous admission of the error of his cosmological constant. However, in the last 15 years astronomers have found that they may have to reconsider the cosmological constant. In 1998 two teams announced the result of attempts to measure the expansion rate of the universe in the past. Everyone expected the universe's expansion to be slowing down; after all gravity is an attractive force, so it should be slowing down the outward motion of the galaxies from the big bang as time goes on. But they found that the expansion of the universe is speeding up, at a rate faster now than it was when it was half its present age. A mysterious dark energy – a force that seems to be a property of space itself – has the effect of pushing galaxies apart at an ever-increasing rate. The nature of dark energy is still unknown, and one possibility is that it is Einstein's cosmological constant. Further investigation is required to determine whether the dark energy has changed with time. Since the initial discovery in 1998, astronomers have been able to look further back to a time when the universe was expanding more slowly. We currently seem to be living in an era in which dark energy dominates the gravitational pull between clusters of galaxies, but in the past this does not seem to have been the case. Initially the expansion of the universe was slowing down, but then the dark energy began to dominate and the expansion rate increased. If dark energy is a property of space then this would be expected, as when there is more space the amount of dark energy should increase.

What Then?

If the universe's rate of expansion increases as the universe gets bigger, the universe will expand faster and faster – this is sometimes referred to as 'the big rip'. In the end galaxies will be moving apart so fast that we won't see their light. Particle physics predicts a cosmological constant, but it is some 120 orders of magnitude greater than the one observed– one of the many puzzles related to dark energy.

What Gives?

75 Percentage of the universe that is dark energy under the current 'lambda dark matter' model of the contents of the universe.

4 Percentage of the universe that is 'normal' matter under this model; 21 per cent is thought to be dark matter.

72 km/sec per Mega parsec The universe's current rate of expansion is measured in terms of the Hubble constant. Its value would have been less when the universe was about half its current age.

What Else?

What if antimatter felt antigravity? *See page 62*

What if the universe was infinite? *See page 90*

WHAT IF THERE WERE PARALLEL UNIVERSES?

Sophie Hebden

Many scientists believe that our universe exists alongside a vast number of universes, which are together called the 'multiverse'. This would be impossible to prove, because we can't observe anything outside our own universe, but various theories predict the multiverse – whether as expanding bubbles of spacetime that arise out of chaotic inflation, or instabilities from string theory or from the 'many worlds' interpretation of quantum mechanics. In the multiverse, each different universe could have different numbers controlling the laws of physics, such as the speed of light, or the charge of an electron. The numbers we see in our historical universe – known as the fundamental constants – are finely tuned, determining the different particles and forces that make up reality. But no one knows why they take the precise values they do. For example, if you tinker with the fine structure constant, which governs the strength of the electromagnetic force, making it 4 per cent bigger, carbon would not have been created in stars; 4 per cent smaller and we would have no oxygen. Either way, life as we know it could not exist. The multiverse gets around the fine-tuning mystery because if there are a vast number of possible configurations of the fundamental constants, at least one will spring into existence that gives rise to Nature as we see it. This is an explanation for the anthropic principle, which says that the universe has to be the way it is, or we would not be here to comment on it. Another mystery the multiverse could resolve is why entropy in the universe, or disorder, always increases. Spilt coffee never returns itself to the cup. But perhaps the only reason why certain bubble universes are born is because they start from a low-entropy state. Other work suggests our universe began differently, with a good deal of entropy, and passed into a different bubble by a sort of quantum tunnelling process, akin to the way quantum particles can jump from one place to another.

What Then?

We might be able to find evidence for the other bubble universes because they could collide with our universe in a 'bubble collision'. This could leave a mark on the cosmic microwave background radiation – the leftover radiation from the big bang – which might look like a circular patch of different temperature. Nothing conclusive has been found, and no one really knows what happens in a bubble collision: perhaps, as with soap bubbles, there'll be a big 'pop' and it will all vanish.

What Gives?

100,000 – & add another 495 zeroes Number of solutions to the equations of string theory. If you had a bubble of spacetime to match each solution, this is how many bubbles there would be.

1895 Year in which the word 'multiverse' was first used – by American philosopher and psychologist William James.

What Else?

What if the big bang theory were wrong? *See page 82*

What if everything were made of string? *See page 84*

WHAT IF THE BIG BANG THEORY WERE WRONG?

Rhodri Evans

Until the 1960s the most popular theory of cosmology was the 'steady state theory', according to which new matter is constantly emerging in a continuously expanding universe. Its main proponent was English astronomer/mathematician Fred Hoyle. But there is now solid evidence for a big bang 13.7 billion years ago. The cosmic microwave background radiation is evidence of this hot, early universe. Other signs include the observed amount of hydrogen, helium, lithium and beryllium in the universe. But was this event 13.7 billion years ago really the beginning of the universe? It need not have been, nor is our universe necessarily the only one. A group of cosmological models, collectively known as cyclic cosmological models, suggest that our universe may be undergoing an expansion phase that is one in an infinite series of expansions and contractions. In 2010 English mathematical physicist Roger Penrose proposed the conformal cyclic cosmology model, in which the universe undergoes an infinite series of expansions – when all the matter has been converted to radiation, a new 'big bang' occurs. He asks whether concentric circles found in our image of the cosmic microwave background are evidence of an earlier universe before the current big bang. In addition, physicists are paying more attention to the idea of the multiverse, which suggests that our universe is but one of many. M-brane theory is a branch of the science that investigates this. The idea is that multiple universes exist in separate membranes (branes), and for most of the time each is unaware that others exist. But now and then they may collide, and if they did, such a collision should leave some kind of evidence. If we are in the latest cycle of an infinite series of expansions and contractions, the question arises as to whether each expansion phase looks the same as the last. The answer is probably not, because there are sufficient random events in the early universe for the universe to be quite different each time it develops.

What Then?

One of the most puzzling questions in cosmology is 'Why does the universe seem so perfectly fine-tuned to give rise to stars, planets, life and human beings?' If there is an infinite number of cycles in the prehistory of our universe, or if the multiverse theory is correct, much of this fine-tuning can be attributed to chance.

What Gives?

379,000 Approximate age of the universe in years when the cosmic microwave background radiation began.

2.725K Equivalent temperature of the cosmic microwave background radiation. This has cooled over time with the expansion of the universe from an original 3,000K. (2.725K is –270.425°C or –454.765°F; 3,000 K is 2,726.85°C or 4,940.33°F.)

What Else?

What if there were parallel universes? *See page 80*

What if everything were made of string? *See page 84*

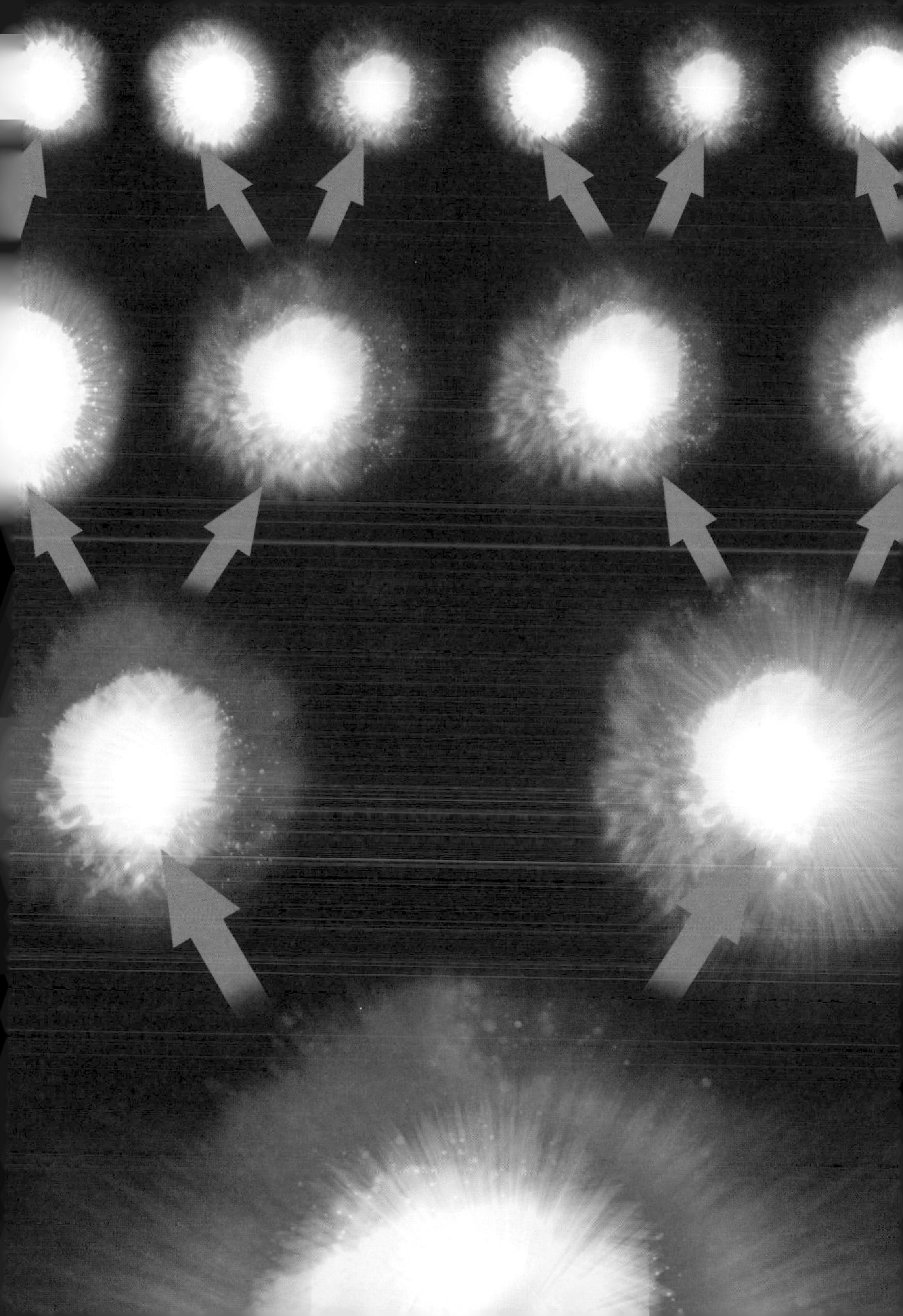

WHAT IF EVERYTHING WERE MADE OF STRING?

Brian Clegg

When the ancient Greeks first came up with atomic theory, the idea was to simplify the structure of everything in the cosmos to its most basic components – and when modern atomic theory was developed it seemed that things were indeed very simple. Yet over time the number of particles required to make the model work has grown significantly. Add in other particles acting as carriers for the various forces and the resultant 'standard model' is complex indeed. However, there is one possibility that returns to a single 'atom' – and could even offer alternatives to cosmological theories like the big bang. All that's necessary is to base our universe on string. This is not literal string, but the idea that each particle is the result of a truly fundamental entity – a 'string', which vibrates different ways. String theory has a distinctly ancient Greek feel about it, because instead of being deduced from observations of the universe around us, it is a possibility that emerges from pure mathematics, which has then been fitted to the world. Perhaps the biggest obstacle to the theory is that unlike our familiar four dimensions (three of space and one of time), string theory requires there to be ten. The extra six spatial dimensions are clearly not part of our everyday experience. To get around this, it is assumed that the dimensions are curled up so tight that they are not detectable. There are several variants of string theory, which have been pulled together to form M-theory, adding yet one further spatial dimension and with a basic unit of a 'brane', a multidimensional membrane that in its simplest, single-dimensional form is a string. In M-theory our universe is thought to be a three-dimensional brane, floating through higher dimensions of space. If M-theory is true it provides alternatives to the big bang – in which, for instance, the universe as we know it began with the collision of two branes.

What Then?

Hundreds of physicists have worked on string theory, but it does have one major flaw – it lacks any significant testable prediction that can be checked against reality. Some physicists have called it 'not a theory, just a hunch' or 'not even wrong'. German physicist Martin Bojowald, a supporter of the alternative loop quantum gravity theory, has pointed out that string theory, with its inability to make predictions, is the ultimate theory of everything because everything (and anything) can happen in it.

What Gives?

10^{80} Number of protons in the universe. There are more possible solutions to the equations of string theory than there are thought to be protons in the universe.

9 Number of spatial dimensions in string theory.

5 Number of incompatible variants of string theory brought together in M-theory.

What Else?

What if we had GUTs and TOEs? *See page 58*

What if space and time came in loops? *See page 86*

WHAT IF SPACE AND TIME CAME IN LOOPS?

Sophie Hebden

String theory isn't the only game in town when it comes to constructing a theory of quantum gravity. One of the biggest open questions in physics is how we can merge general relativity – which describes gravity as a geometric property of space and time – with the quantum physics of the atomic realm. Loop quantum gravity (LQG) is an increasingly popular approach. In LQG, space has a microstructure that consists of a network of discrete edges joined by nodes, like on an aeroplane route map. This structure is called a spin network. The links between nodes can wrap around one another to form loops, or braids, whose twists and turns endow the different fundamental particles with their particular properties. For example, an electron can be formed from a deformed pretzel-like braid, but if you add three anticlockwise twists to it, you have a positron. This works for the lightest particles but there are unsolved problems in working it out for the heaviest particles to date. Work on LQG started in the mid-1980s, and is based on a reformulation of general relativity by Indian physicist Abhay Ashtekar, which brings the mathematical language closer to that used in particle and quantum physics. One of its most important successes is predicting the exact entropy or information content of black holes and the radiation they emit. These were key theoretical achievements developed by Israeli physicist Jacob Bekenstein and his British counterpart Stephen Hawking in the 1970s and early 1980s. That LQG can emulate this result is crucial because it brings together the physics of gravity with the realm of thermodynamics and information. Using LQG to describe a black hole is interesting because instead of getting a mathematical singularity (a point of infinite density) at its core, the black hole opens out into another region of spacetime. If you apply it to the big bang, LQG predicts an eternal universe.

What Then?

Einstein's special theory of relativity maintains that the speed of light is a universal constant, independent of motion. But astronomers have found that radiation from gamma-ray bursts (produced in explosions billions of light years away) doesn't arrive all at the same time. It may be possible to distinguish whether high-energy photons arrive later because of how they are produced in the explosion, or as a result of an accumulated effect during their journey across the universe, perhaps due to the microstructure of spacetime predicted by LQG.

What Gives?

Spin-foam theory Variant of LQG that describes the quantum geometry of spacetime.

10^{-35} m Size of the indivisible chunks of space predicted by LQG.

What Else?

What if we had GUTs and TOEs? *See page 62*

What if everything were made of string? *See page 84*

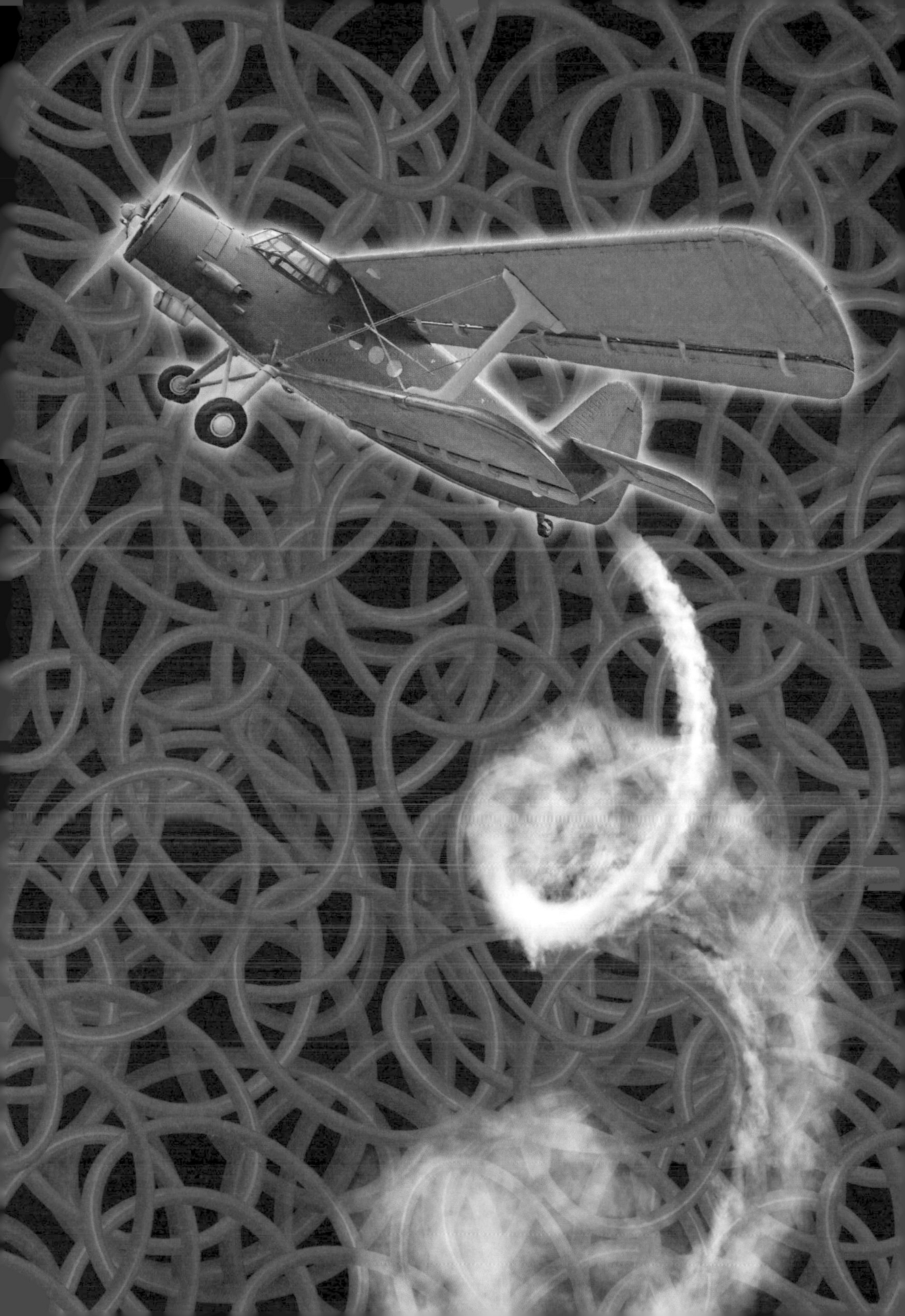

HISTORICAL

WHAT IF THE MILKY WAY IS NOT THE UNIVERSE?

At the beginning of the 20th century most astronomers thought that our galaxy, the Milky Way, was the only stellar system and that our Sun lay at its centre. This viewpoint was best summarized by the so-called Kapteyn Universe, named after the Dutch astronomer Jacobus Kapteyn who made detailed observations of the number of stars per unit area in the disc of the Milky Way. However, in the 1910s American astronomer Harlow Shapley – who was studying Cepheid variables (a group of highly luminous stars that vary in brightness over time) – used the recently discovered relationship between the stars' brightness and length of pulsation to show that the distribution of globular clusters (spherical groups of stars) as seen from Earth was not consistent with our being at the centre of the Milky Way, but that we are instead out in the disc of a flat distribution of stars that orbit the centre. In the same decade another American astronomer, Heber Curtis, argued that the number of observed novae (new stars) in the Andromeda nebula seemed to be greater than the number in the Milky Way galaxy, which he felt was unlikely if the Andromeda nebula were part of the Milky Way.

On 26 April 1920 these two astronomers came together at the Smithsonian Museum of Natural History in Washington, DC, USA, to discuss their disparate views of the nature of our Milky Way galaxy and of the universe. This event has come to be known as the Great Debate. They argued three points: the Sun's position in the Milky Way, the size of the Milky Way and whether 'spiral nebulae' were part of the Milky Way galaxy.

Shapley argued that our Milky Way galaxy was the entire universe, but Curtis believed that the Andromeda and other similar nebulae were – like our own – 'island universes' (a term first coined by the German philosopher Immanuel Kant to refer to what we would now call galaxies). The matter was settled in 1923 by observations made by American astronomer Edwin Hubble using the 254-cm (100-in) telescope at the Mount Wilson Observatory near Los Angeles, California, USA.

Hubble was observing the Andromeda nebula, and identified a number of novae in the system. However, because he was making regular observations he noticed that one of these objects was not a nova after all, but a Cepheid variable. Because it had been shown that Cepheid variables vary their light with a period related to their intrinsic brightness, these stars could be used to estimate interstellar distances and Hubble was able to use this Cepheid variable in Andromeda to show that the Andromeda nebula was far too far away to be part of our Milky Way galaxy. Once Hubble had shown that the Andromeda nebula was in fact a galaxy in its own right, it became easier to argue that similar spiral nebulae were also galaxies external to ours.

After discovering the nature of these spiral nebulae as spiral galaxies, Hubble went on to measure their spectra (the range of frequencies of the light they gave off) and found that nearly all of them were shifted towards the red, showing that they seemed to be moving away from our galaxy. Not only this, he also found their speed of recession was in proportion to their distance from us. He announced this discovery in 1929. He had discovered the expansion of the universe. A natural consequence of this observed expansion is that the universe would have been smaller in the past, and hence must have had a finite beginning in time – the big bang.

WHAT IF THE UNIVERSE WERE INFINITE?

Rhodri Evans

How big is the universe? The simple answer is that we do not know. We have strong evidence that the universe began 13.7 billion years ago in a very hot, dense state, and that it has been expanding ever since. We can see galaxies back as far as 13 billion years ago, when the universe was less than 1 billion years old. And the oldest radiation we can see – the cosmic microwave background, the echo of the big bang – comes from the time, about 400,000 years after the big bang, when the universe had cooled enough for the hydrogen in it to become neutral rather than being ionized. If the universe is 13.7 billion years old, we can only see light that has been travelling for 13.7 billion years, although because of its expansion the observable universe is significantly bigger than 27.4 billion light years across. It may be that the universe is very much bigger than a sphere with this radius, maybe many thousands of times bigger. It may even be infinite. We do not know, because we can only see as far as the distance light can travel in 13.7 billion years. Einstein's general theory of relativity linked gravity to geometry, and so cosmologists talk about the geometry of the universe. A flat universe is one that has what we call the critical density, a density just low enough for there not to be enough matter to stop it from collapsing again. All measurements of the geometry of the universe so far suggest that its geometry is indeed flat. But it could be that we see such a tiny part of the universe that we are only measuring the local geometry. We all know the surface of the Earth is curved, but if you measure the curvature of the Earth's surface in, say, your back garden, you will not detect any curvature because the part you are looking at is too small. The same may apply to our measurements of the geometry of the universe.

What Then?

In an infinite universe there would be infinite possibilities, including a probability of more than zero that somewhere else there is a carbon copy of our planet Earth – and even a carbon copy of you and me. If the universe was infinite, then irrespective of how much matter was in it, it would carry on expanding forever, because it occupies infinite space and so its density tends towards zero.

What Gives?

93 billion light years Distance from side to side of the observable universe, as it has expanded during the time the earliest light has been heading towards us.

10^{78} Times the volume of the universe is thought to have increased during the inflationary period in the first fraction of a second of its existence.

What Else?

What if Einstein was wrong? *See page 78*

What if there were parallel universes? *See page 80*

WHAT IF ANTIMATTER WERE EVERYWHERE?

Brian Clegg

Antimatter may sound like something out of science fiction, but it's a real physical phenomenon. It's like ordinary matter in most ways, except that antimatter particles have the opposite electrical charge to matter particles. The first antimatter particle predicted to exist was the positron or antielectron, like an electron but with a positive charge. Even uncharged particles like the neutron have their antimatter equivalent. Although the antineutron has no charge, the quarks that make it up have opposite charges to those in a conventional neutron. Where a neutron is made up of a positively charged up quark and two negatively charged down quarks, an antineutron consists of a negatively charged up antiquark and two positively charged down antiquarks. We don't see antimatter lying around because equivalent matter and antimatter particles annihilate each other, converting their mass into a considerable amount of energy, determined by $E=mc^2$, where 'c' is the formidably large speed of light. The Enterprise in American TV show *Star Trek* is powered by an antimatter reaction. Although the mechanism is fictional, the idea makes sense: the matter/antimatter reaction is the most concentrated form of energy there is. According to the big bang theory, the universe began as a seething ball of energy. As it expanded and cooled, this energy would convert into matter and antimatter in equal quantities. The universe as we see it consists almost entirely of conventional matter. The best supported idea of why this is the case is that there may have been a subtle violation in the expected symmetry between particles and antiparticles under conditions that would mean there would be slightly more matter particles than antimatter. The rest would annihilate, leaving the matter we see today. Alternatively, it is possible there are huge regions of antimatter outside the part of the universe we can see.

What Then?

We can make antimatter – but only in tiny quantities. If we had some way to access naturally occurring antimatter it would provide an intense source of energy – but it is very difficult to store. Charged antimatter particles like positrons and antiprotons can be kept away from matter by holding them in an electromagnetic trap, but this can't be done with neutral particles and complete antiatoms, although the small magnetic field caused by particle spins gives a slight ability to manipulate them.

What Gives?

1 millionth of a gram The typical amount of antimatter currently produced worldwide in a year.

10 years The amount of time it would take a typical power station to produce the amount of energy generated by combining 1 kg (2.2 lb) of matter and antimatter.

What Else?

What if antimatter felt antigravity? *See page 62*

What if empty space were full? *See page 106*

ASTROP

ASTROPHYSICS

ASTROPHYSICS INTRODUCTION

Where cosmology tells us about the origins and nature of the universe, astrophysics explores the workings of stars and other inhabitants of space.

For many years, a mystery that faced scientists was how stars like the Sun kept burning for so long. The obvious assumption was that they were on fire. There was nothing else known that produced a similar combination of heat and light. But there was a problem. Calculations based on a Sun made of coal – the only familiar substance that burned so consistently – suggested it would only last a few million years. Yet by the 19th century there was good evidence that the Earth (and hence the Sun that formed it) had been around much longer than that.

Early geological estimates put the age of the Earth at 1 billion years, a figure that gradually increased to the current estimate of 4.5 billion. Nothing could burn for that long. It was only when nuclear reactions were discovered that it became clear there was another potential power source for the Sun – nuclear fusion. Not the breaking apart of chemical bonds in burning, but hydrogen nuclei fusing to form helium, the next heaviest element, and releasing energy in the process.

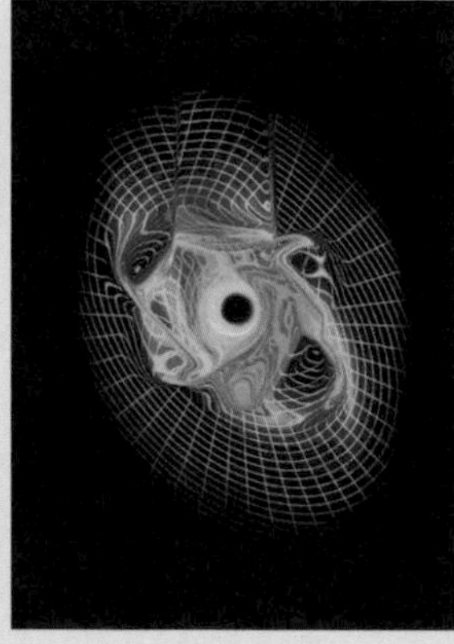

We can tell that these and other chemical elements are present because each element absorbs different frequencies of light, leaving black bars in the spectrum of light produced by the star, acting as a kind of optical fingerprint.

Astrophysics took a huge leap forward with Einstein's general relativity. For the first time we had an explanation for gravity and its interaction with the other forces acting on the particles that made up a star. It was predicted that there should be immensely dense stars – neutron stars – left over in the aftermath of a supernova, where a star has exploded, suddenly producing a burst of light that can be brighter than a whole galaxy. General relativity also suggested that there could be an ultimate collapse of some stars, where everything in the star was reduced to a point in spacetime. Even light could no longer escape from such an exotic body, which in the 1960s was given the name 'black hole'.

The universe has continued to produce surprises, from the discovery of pulsars with their fast-repeating radio signals, to quasars, now thought to be incredibly bright baby galaxies. For the astrophysicist, space is an exotic zoo with inhabitants that will always fascinate.

WHAT IF
THERE WERE NO TIDES ON THE MOON?

Rhodri Evans

English physicist Isaac Newton was the first to explain the tides successfully. He showed, from his law of gravity, that the gravitational influence of the Moon would produce a bulge on both sides of the Earth, leading to two high and two low tides each day. In fact, in astrophysics the term 'tide' has a more general meaning: it is the different gravitational force felt by different parts of an extended body. The side of the Earth nearer the Moon feels a different gravitational force to the Earth's centre, which feels a different force to the side of the Earth furthest from the Moon. Newton also stated that to every action there is an equal and opposite reaction. Just as the Moon produces tidal forces on the Earth, the Earth produces tidal forces on the Moon. One consequence of the tidal force the Earth produces on the Moon is that the Moon's rotation on its axis is tidally locked to be the same as the time it takes to orbit the Earth. This means that we always see the same face of the Moon; it takes 27.3 days to orbit the Earth, but it also takes 27.3 days to rotate once on its axis. If one were to stand in a particular place on the surface of the Moon with the Earth visible in the sky, the Earth would stay at the same height relative to your horizons all the time. The tidal forces produced by the Moon cause both the water and land on the Earth to rise and fall twice a day, but the water deforms more because it is liquid. We are all familiar with the change in the height of the sea on a beach between high and low tide. As the water flows in and out between high and low tide the motion of the water can be used to generate electricity. There are numerous projects afoot to tap into this regular, reliable, daily source of renewable energy to generate electricity.

What Then?

The tidal force of the Moon on the Earth is slowing down the Earth's rotation. We have evidence from the growth rings of trees that in the past there were more days in a year – that is, that the days were shorter. To conserve angular moment, as the Earth's rotation slows down the Moon is moving further from the Earth, about 3 cm (1.2 in) per year. We are able to confirm this motion by firing lasers that reflect off mirrors left on the Moon's surface by the astronauts of the NASA *Apollo* missions.

What Gives?

9 billion watts
Amount of electricity that would be generated by tidal movement of water if a 15-km (9-mile) barrage were built across the Severn Estuary in southern England – more than 5 per cent of the UK's total energy needs.

50 Number of Earth days that makes a single day on Mercury. The tidal force produced by the Sun on Mercury locks its rotation. Mercury rotates three times for every two orbits of the Sun.

What Else?

What if gravity is not a force?
See page 48

What if black holes had hair?
See page 110

WHAT IF A SPOONFUL OF MATTER WEIGHED 100 MILLION TONNES?

Rhodri Evans

What has the mass of the Sun but is the size of a city? The answer is a neutron star, a remarkable object in which all the space in atoms has been squeezed out. Most of an atom is empty space. The nucleus, where virtually all the mass exists, is just a tiny part of the size of an atom. If the nucleus were the size of a grain of rice, an atom would be about the size of a football stadium, with nearly all of that volume just being empty space. In a neutron star, that void has been squeezed out. In the early 1930s Indian theoretical astrophysicist Subrahmanyan Chandrasekhar calculated that if the stellar remnant we call a white dwarf had a mass more than 1.4 times that of the Sun, gravity would squeeze all the space out of atoms and the white dwarf would collapse to become a sphere of pure neutrons, with no space between them. The electrons in the atoms would be forced into the nucleus in which they would combine with the protons to produce a sphere of pure neutrons (a process called reverse beta decay). All the space in atoms would be squeezed out by this collapse. This is what we call a neutron star. Our own Sun is not massive enough to end its life as a neutron star, but for a star whose initial mass is about three times the mass of the Sun, its stellar remnant will be too massive to support itself as a white dwarf. It will collapse to become a neutron star. Neutron stars are virtually the most dense objects we know about. The only objects we know of that are more dense than neutron stars are black holes. If a man weighing 75 kg (165 lb) here on Earth were to stand on the surface of a neutron star, he would weigh more than 10 billion tonnes (11 billion tons)!

What Then?

When white dwarfs collapse into neutron stars they emit a huge flux of neutrinos, possibly the most elusive particles in Nature. These neutrinos can pass through the Earth unimpeded. Detecting a flux of neutrinos from space gives us our first indication of the collapse of a white dwarf into a neutron star. When the surrounding material of the star collapses onto the neutron star it rebounds and the star explodes in a supernova. It is in this supernova explosion that the elements above carbon in the periodic table are created.

What Gives?

1054 Year in which a neutron star was formed at the centre of what we now call the Crab Nebula. The supernova explosion was visible in the daytime for over one month. When a star collapses to form a neutron star it briefly becomes brighter than the galaxy within which it resides.

40 billion:1 The ratio of a neutron star's density to that of lead.

What Else?

What if a star could rotate 600 times a second? *See page 102*

What if stars and supernovas make the elements? *See page 104*

WHAT IF A STAR COULD ROTATE 600 TIMES A SECOND?

Rhodri Evans

In November 1967 British PhD student Jocelyn Bell wrote down 'LGM' on a plot of a radio signal. 'LGM' stood for 'little green men'. For a brief period, this seemed to be the best explanation of the radio signal Bell and her supervisor, British radio astronomer Antony Hewish, had discovered. It was a regular radio pulse, each of the pulses arriving exactly 1.33 seconds apart. So regular were the radio signals that Bell and Hewish initially thought that a signal from an extraterrestrial civilization was the most likely explanation. It was not until another regular radio signal was discovered in an entirely different part of the sky that the 'LGM' theory was abandoned. What Bell and Hewish had in fact found was the radio signal from a rapidly rotating neutron star. The term 'pulsar' was coined to describe such objects, and a theory was quickly developed to explain them. Neutron stars have intense magnetic fields, and this magnetic field can rip electrons from the surface of the star. As the electrons travel away from the magnetic poles of the neutron star they are accelerated to very high speeds and emit radio waves. It is the rotation of the neutron star that produces the pulsing; we are seeing the beam of radio radiation sweeping across our line of sight as the neutron star rotates, just like the beam of light from a lighthouse. In 1968 a pulsar was discovered at the centre of the Crab Nebula, known to be the remnant of a supernova that exploded in 1054. The Crab Nebula pulsar has a period of only 33 milliseconds, so the neutron star is rotating more than 30 times a second. In 1982 a pulsar was discovered with a period of just 1.6 milliseconds, meaning the star is rotating 625 times a second.

What Then?

Many pulsars have such accurate and short periods that they can rival our best atomic clocks for keeping time. We have measured the slowdown of pulsars in binary systems, and can use these measurements as an indirect test of a prediction of general relativity – the existence of gravitational waves. The first extrasolar planet to be discovered, in 1994, was discovered orbiting a pulsar. Regular variations in the period of the pulsar indicated that it was being orbited by an unseen object.

What Gives?

10 km Typical radius of a neutron star. This volume contains more than the mass of our entire Sun. (10 km is 6.2 miles.)

130 billion Number of times the force of gravity on the surface of a neutron star with a 10-km (6.2-mile) radius and a mass of 1.5 times the mass of the Sun is stronger than the force of gravity at the Earth's surface.

What Else?

What if gravity is not a force?
See page 48

What if a spoonful of matter weighed 100 million tonnes?
See page 100

HISTORICAL
WHAT IF STARS AND SUPERNOVAS MAKE THE ELEMENTS?

In the 1920s English astrophysicist Arthur Eddington realized that stars get their energy from converting hydrogen to helium in their cores. In fact, stars and the universe in general are nearly entirely hydrogen (about 74 per cent) and helium (about 24 per cent); the other elements comprise less than 2 per cent. Yet the Earth's crust is about 46 per cent oxygen, 28 per cent silicon, 8 per cent aluminium and 5 per cent iron, while the human body is about 65 per cent oxygen, 18 per cent carbon and 3 per cent hydrogen. Clearly the composition of the stars and universe is very different to that of the Earth or our bodies. From where do these other elements come? And how were hydrogen and helium formed in the first place?

Scientists realized in the 1940s that the early universe would have been much smaller, denser and hotter than the present universe. This work, led by Russian-American George Gamow, is called 'big bang nucleosynthesis'. In the early universe the first element to form was hydrogen, the simplest element. Within the first three minutes of the universe's lifetime, it was hot and dense enough for four hydrogen atoms to come together to form helium. But, as time went on the universe was expanding and getting cooler and less dense, so after three minutes the synthesis stopped, giving us our ratio of hydrogen and helium. Thus big bang nucleosynthesis was able to explain the observed abundances of hydrogen and helium. But where did all the other, heavier elements come from?

English astronomers Fred Hoyle, Alfred Fowler and Margaret and Geoffrey Burbidge showed in the 1950s that the heavier elements were formed inside stars and in supernova explosions. This is called 'stellar nucleosynthesis'. Relatively low-mass stars like our Sun will burn hydrogen in their cores to produce helium, and will spend some 90 per cent of their lifetimes doing this. Once all the hydrogen in the core is used up, the star expands into a red giant by burning hydrogen in a shell around the core. The core becomes hot enough for helium to be burned, producing carbon. A solar-mass star cannot get beyond this point: the temperatures and pressures in the core are not high enough. Such a low-mass star ends its life by blowing off its outer layers in a planetary nebula, leaving behind a sphere of carbon that we call a white dwarf.

However, more massive stars (more than about five times the mass of the Sun) will have higher temperatures and pressures in their cores, and so are able to burn carbon to produce oxygen, burn oxygen to produce silicon – and so on all the way up to iron. Then because iron is the most tightly bound nucleus, no energy can be released in burning iron, so the process stops. The sudden extinguishing of the nuclear furnace causes the surrounding layers to collapse onto the new iron core, rebound and rip the star apart in what we call a supernova, the explosion of a high-mass star. It is in the supernova explosion that the elements higher than iron are produced, all the way up to uranium. Thus our bodies, comprising mainly oxygen and carbon, are very much star dust. We have the stars to thank for our existence.

WHAT IF EMPTY SPACE WERE FULL?

Frank Close

No one has succeeded in making nothing. The vacuum of space above the Earth's atmosphere is actually seething with stuff, some familiar, some bizarre. Hundreds of years ago people believed that once you made a vacuum, by removing all the air and gases, you were left with truly empty space. Yet the vacuum is weird. We can see the Sun but cannot hear it. Sound waves require a medium and so the absence of air hides the Sun's violence from our ears, which is just as well: the shaking would probably destroy all solid matter on Earth. We can see the Sun because light – electromagnetic waves – can travel across space. But light consists of waves. In what medium are these waves? Surely you need a medium to make a wave? In the late 19th century this made people think that space is full of some essence, which they called 'ether'. Unfortunately this concept did not fit with other facts, for instance that planets travel through space without any ethereal impediment. Einstein's special theory of relativity did away with ether, and today we visualize space as full of electric and magnetic fields, which can wobble, along with gravitational fields, too. It is far from empty. Quantum theory also tells us that space is seething with transient 'virtual' particles of matter and antimatter that flit in and out of existence. The more powerful the microscope with which you examine 'empty' space, the more violent this quantum foam appears to be. These ideas explain many phenomena, such as the way that the fundamental forces between particles behave at very short distances, and under suitable circumstances these virtual particles can be ejected from the vacuum and studied. The discovery of the Higgs boson suggests that space is also filled with a further field – the Higgs field – that gives mass to fundamental particles, such as the electron. We are immersed in an ether after all, but one that satisfies the constraints of relativity theory.

What Then?

Space is filled with gravitational fields. Therefore there ought to be gravitational waves, too. So far no evidence has been found, but this is probably because the necessary experiments are very difficult. Although the Higgs field is known to exist, we need to know how it is formed, what its structure is and whether it also contains novel virtual particles. Dark energy fills space, too, and we know nothing about the stuff of which it consists.

What Gives?

1643 Year in which Italian physicist Evangelista Torricelli made a vacuum.

Lamb shift Subtle effect in the spectrum of hydrogen, due to virtual particles affecting the motion of the electron in the hydrogen atom.

What Else?

What if the Higgs boson did not exist? *See page 66*

What if antimatter were everywhere? *See page 92*

WHAT IF WE DUG A BLACK HOLE?

Rhodri Evans

Black holes are some of the most mysterious objects in the universe. Although we tend to associate them with Einstein's general theory of relativity, their existence was first suggested as far back as 1783 by English geologist John Michell. However, it is only with Einstein's theory that we have gained a proper understanding of some of the very weird properties of black holes. They are not simply a theoretical idea; strong evidence for their existence first emerged in the early 1970s with observations of Cygnus X1, the strongest X-ray source in the constellation Cygnus. Material falling into a black hole becomes heated to millions of degrees and emits X-rays. Put simply, a black hole is an object so dense that not even light can escape from it. With our current understanding, the density at a black hole's centre is infinite – but this simply shows that our current theories cannot fully explain the extreme physics found there. If we were to approach a black hole in a spaceship we would find that the speed needed to escape from its gravitational pull would get larger until a point came at which it was equal to the speed of light. This is the event horizon of a black hole, the point of no return. Due to the effect general relativity has on time, at the event horizon of a black hole time can appear to stand still. If you observed the spaceship from afar you would see it frozen in time at the event horizon. The hapless travellers in the spaceship would be torn apart by the tidal effects of the black hole's intense gravitational field. We have known since the early 1980s that our Milky Way galaxy harbours a supermassive black hole at its centre, with a mass millions of times that of the Sun. And the Hubble Space Telescope has found that all galaxies seem to have a central supermassive black hole. There seems to be a strong relationship between the mass of the central black hole and how quickly the stars are moving in the whole galaxy. This suggests black holes play a crucial role in the formation of galaxies.

What Then?

Because our theories predict an infinite density at the centre of a black hole (called a singularity), we know that our current best theory of gravity, the general theory of relativity, is incomplete. A quantum theory of gravity would hopefully remove these embarrassing infinities from our calculations. It's possible that black holes form wormholes, tunnels to other parts of the universe. In theory, you could travel into a black hole and pop out somewhere else in the universe. In practice you'd be torn apart by the black hole's intense gravitational field.

What Gives?

20 mm Diameter of the Earth if it were squashed down to form a black hole. In theory, any object can be a black hole, it just needs to be dense enough. (20 mm is 0.8 in.)

4 million The number of times the supermassive black hole at the heart of the Milky Way galaxy is thought to be more massive than the Sun.

What Else?

What if there were a smallest distance? *See page 32*

What if black holes had hair? *See page 110*

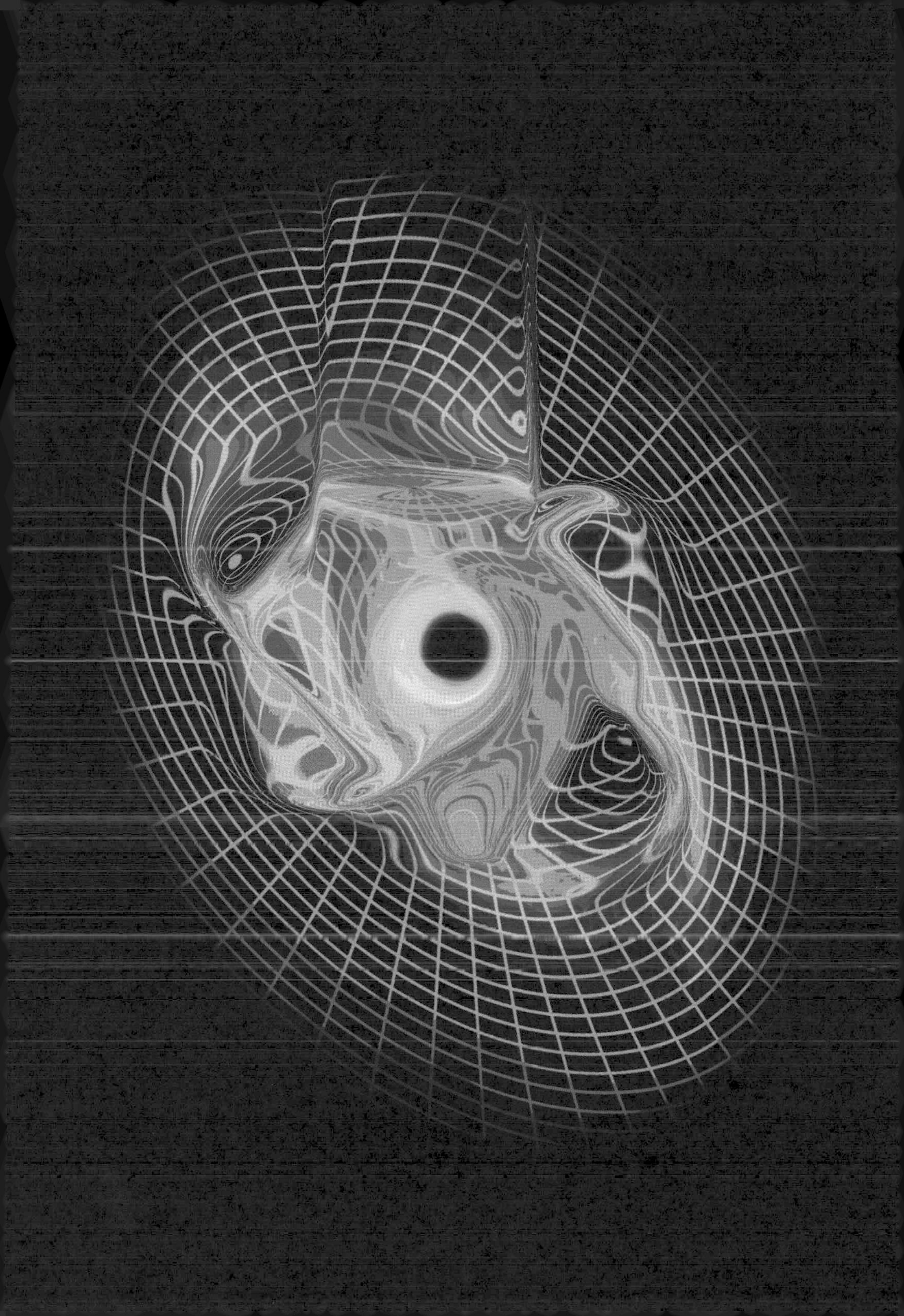

WHAT IF BLACK HOLES HAD HAIR?

Brian Clegg

Black holes are the strangest phenomena in the universe – and their image of being bizarre and wonderful is, if anything, magnified by the existence of the 'no hair theorem'. Before diving into stellar coiffure, it is worth getting a better idea of just what happens at the point where everything changes for a black hole. The black hole itself has no spatial dimensions – it is a point – but the event horizon is a sphere that defines the 'size' of the black hole as seen from the outside. It is the horizon of no return – the distance from the black hole where nothing, not even light, can escape. If you took a trip into a black hole, the chances are you would not survive long enough to notice that you were crossing the event horizon. This is because the gravitational pull would increase with a very steep gradient as you got closer and closer to the collapsed star that had formed the black hole. In fact the gradient would be so steep that (assuming you were travelling feet first) the pull on your feet would be much greater than the pull on your head. This difference in gravitational pull – a tidal effect – would mean that you would be stretched out to become a long, thin cylinder, a process that cosmologists, with a rare burst of humour, give the technical term 'spaghettification'. Relativity comes into play, too. General relativity predicts that as you get closer to the event horizon an outside observer will see time slow down for you until it effectively comes to a stop at the horizon – but you would not notice this. The 'hairless' aspect of black holes reflects the idea that we can never know anything more about what's in a black hole than its mass, angular momentum and charge because nothing would ever be emitted. But Stephen Hawking has predicted that some black holes would produce a constant stream of particles, so-called Hawking radiation.

What Then?

A new black hole is likely indirectly to produce a great deal of radiation as a result of accelerating nearby particles as they rush into the hole, but Hawking radiation is a quantum effect. Quantum theory predicts that in empty space, pairs of matter and antimatter particles are constantly popping in and out of existence. If this happens at the event horizon, one could be sucked into the black hole while the other zooms off, producing the black hole's quantum 'hair'.

What Gives?

3 solar masses
The approximate mass at which a white dwarf star would not collapse into a neutron star but would form a black hole.

9 km The approximate Schwarzschild radius – the radius of the event horizon of a black hole formed from a star of 3 solar masses. (9 km is 5.6 miles.)

What Else?

What if gravity is not a force?
See page 48

What if we dug a black hole?
See page 108

WHAT IF THERE WERE NO DARK MATTER?

Rhodri Evans

Einstein's theory of gravity has been exquisitely well tested on the scale of the solar system, and on those scales has been shown to be in exact agreement with observations. But as we go to larger and larger scales problems arise. At the scale of a galaxy there seems to be more matter than we can see based on how the galaxies are rotating. On the scale of clusters of galaxies the problem is even worse. These are some of the arguments used to infer the existence of dark matter. The search for dark matter is gathering pace, with experiments starting to try to detect the elusive WIMPs (weakly interacting massive particles) thought to be responsible for this mysterious constituent of matter. But is there another possibility: that dark matter doesn't actually exist? All of the evidence for dark matter is gravitational, based on finding discrepancies between the amount of matter we can see with our telescopes and the amount of matter inferred assuming Einstein's general theory of relativity to be correct. But could it not be possible that Einstein's theory is simply not correct on the larger scales? Could it not be the case that at larger distances some modification to his equations are needed? This may indeed be the case, and MOND theories (Modified Newtonian Dynamics) provide an alternative that removes some of the need for dark matter. MOND theories suggest a modification in the strength of gravity when the gravitational force is very weak, when it is about 100 billion times weaker than at the Earth's surface. With this modification, many (but not all) observations can be explained using MOND rather than dark matter. Some problems are actually explained better by MOND than by invoking dark matter, whilst others seem to require dark matter even if MOND is used. It might be that we need both. Of course, should we detect any WIMPs then we will know that dark matter is a reality, although the debate will continue for some time as to how important it is compared to MOND.

What Then?

If there is no dark matter then it would mean one of 20th-century physics' most successful theories, Einstein's general theory of relativity, is not completely correct. But, even if dark matter does exist, it is argued by some that we need MOND anyway. Detecting WIMPs will be a huge vindication of our ability to theorize things that have never been observed before, and is reminiscent of the story of neutrinos. These were predicted by Austrian physicist Wolfgang Pauli in 1930, more than 20 years before they were observed.

What Gives?

80 Percentage of the matter in the universe that is dark matter, if Einstein's general theory of relativity is correct.

1954 Year in which neutrinos were observed.

What Else?

What if space and time came in loops? *See page 86*

What if empty space were full? *See page 106*

2

CLASSI
PHYSI

CLASSICAL PHYSICS

CLASSICAL PHYSICS
INTRODUCTION

It sounds as if classical physics should be the work of the ancient Greeks – and philosophers of the time did write about 'physics', meaning principles of nature. When scientists refer to classical physics, though, they mean the physical theories that were in place by the end of the 19th century, before the seismic shift brought about by relativity and quantum theory.

Some parts of this body of knowledge go back unchanged over centuries. Although the ancient Greeks got most physics wrong, they were trying to answer the questions on which classical physics would focus. For example, the Greeks thought that sight involved light, emitted by a fire in our heads, that shot from our eyes, hit an object and was reflected back. This view was transformed by medieval Arab scientists, who produced a much more modern approach to optics, with light travelling in straight lines from a light source like the Sun, reflecting off an object and coming to our eyes.

By the time of Isaac Newton in the 16th–17th centuries geometric optical diagrams used to describe the progress of light had been formalized, and all that really remained was to settle on

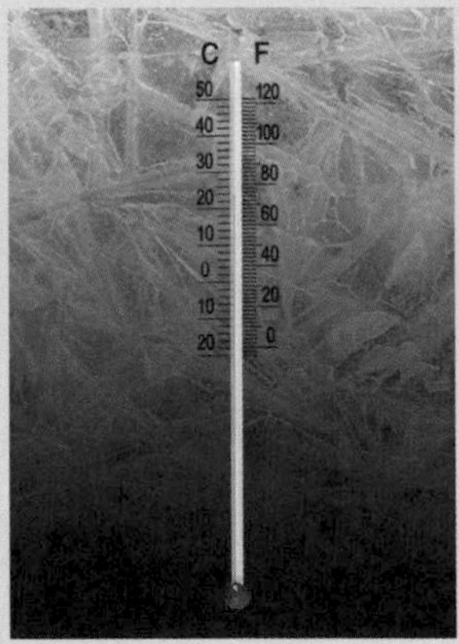

whether light was a particle, as Newton thought, or a wave, as was established by English scientist Thomas Young. This classical view of light still proves useful for designing anything from a pair of spectacles to a giant telescope and is still taught in schools.

Usually classical physics is simpler than the modern equivalent, and produces good enough results to be used for practical, everyday purposes. An example would be Newton's laws of motion. Technically these are a special case of Einstein's relativity. They are approximations that are reasonably accurate when an object is moving slowly compared with the speed of light. This makes the classical approach good enough for everything from engineering to the calculations required to send the *Apollo 11* spacecraft to the Moon.

Classical physics can seem dull alongside the wonders of relativity and quantum theory, but we shouldn't take it for granted. Classical physics is not inherently simple – think, for instance, of fluid dynamics – nor is it without its curiosities. Thermodynamics, for instance, seems plodding enough, describing how heat flows from one place to another. And yet it includes the second law of thermodynamics, which tells us how we can expect the universe to develop – and features an endearing demon, who seems to achieve the impossible.

WHAT IF SOMETHING WERE COLDER THAN ABSOLUTE ZERO?

Simon Flynn

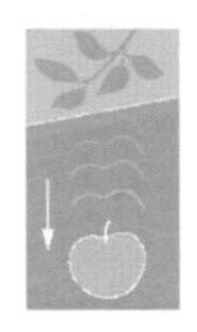

In kinetic theory, the temperature of a substance is a reflection of the motion of the particles in it: the higher the temperature, the faster the average speed of the particles in the matter. So H_2O molecules move faster and vibrate more in steam than in water; more in water than in ice. As temperature is lowered, so is the average motion of the particles. In the 19th century, British physicist William Thomson (Lord Kelvin) realized that this implied a limit to the coldest temperature possible – particles can't move more slowly than when they're not moving at all. He calculated the temperature at which atoms would cease to move to be –273.15ºC (–459.6ºF). This led him to propose a new temperature scale starting from absolute zero and based purely on the laws of thermodynamics. (In the 18th century Swedish astronomer Anders Celsius had based his scale on the freezing and boiling points of water, and Polish-born physicist Daniel Fahrenheit had based his scale on the coldest temperature he could achieve in his laboratory [0ºF or –18ºC] and on his own body temperature [96ºF].) In the Kelvin scale, absolute zero is 0 K, while water freezes at 273.15 K and boils at 373.15 K. The third law of thermodynamics, developed after Kelvin, states: 'A system can't be reduced to absolute zero in a finite number of steps.' This proved to be true. Using a technique known as laser cooling, scientists were able to get agonizingly close to 0 K – a few billionths of a degree from it – but never to the limit. Then scientists at Ludwig Maximilian University in Munich reported in 2013 that they had produced a gas that flipped from a fraction above 0 K to a fraction below: absolute zero had effectively been bypassed. Using lasers and magnets, they induced an ultracold quantum gas of potassium atoms to experience negative pressure that was in turn balanced by negative temperature.

What Then?

The creation of matter with a temperature below absolute zero may provide a clue to dark energy. The universe is expanding at an ever-faster rate in spite of gravitational forces: to explain this, scientists hypothesized the existence of dark energy, which implies strong negative pressure. In the ultracold quantum gas created in Munich, the atoms had to be given negative pressure, making the atoms pull together rather than push apart, but this was balanced by the negative temperature, keeping the gas stable.

What Gives?

2.73 K Average temperature of the universe. (2.73 K = -270.4ºC = –454.8ºF.)

1 K Coldest temperature recorded outside a laboratory. (1 K = –272.2ºC = –457.9ºF.)

What Else?

What if there were a maximum temperature? *See page 122*

What if Maxwell had a demon? *See page 124*

C
F
50
120
40
100
30
80
20
60
10
40
0
20
10
0
20

WHAT IF THERE WERE SUCH A THING AS A FREE LUNCH?

Brian Clegg

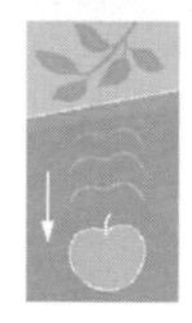

Thermodynamics describes the way heat and energy flow around systems. This was originally the science that explained the workings of the steam engines at the heart of the Industrial Revolution, but over time it has taken on a more central role in our understanding of the workings of the universe. Of the four laws of thermodynamics, the key ones are the first (which describes the conservation of energy, effectively saying energy cannot be created or destroyed) and the second (which can be stated flippantly as 'there's no such thing as a free lunch'). The second law was originally conceived in terms of heat engines, telling us that in a closed system (one in which no energy flows in or out), heat will flow from a hotter body to a colder one. This seems almost too simple to be worthwhile, and yet the second law is crucial to most of the changes that take place in the universe. Another way of stating it is that in a closed system, entropy stays the same or increases. Entropy is a measure of the disorder in a system. The more disorder there is, the higher the entropy. It's easy enough to see how the two versions are related by thinking of a box of gas, divided into two, with hot gas on one side of the division and cold gas on the other. Open the divider, and the gases will mix. Initially the state of the box was quite ordered – with hot, high-speed molecules on one side, and cooler ones on the other. After, there is a confused mix – more disorder and more entropy. It's perfectly possible to reverse the process, but it takes energy to do it. So, for instance, your refrigerator reduces entropy, making a more uniformly cool set of contents, but at the cost of expending energy. If entropy could be reversed without expending energy you could use the process to produce energy – a free energy source or a perpetual motion machine, hence the 'no free lunch' tag.

What Then?

What the early developers of the laws of thermodynamics didn't realize is that the approach is statistical. The second law doesn't really say that entropy will never decrease in a closed system, but that it is statistically unlikely to. There is a small probability that a mix of hot and cold gas will spontaneously separate as each molecule is acting independently – but it is very unlikely in any reasonable timescale.

What Gives?

4 Number of laws of thermodynamics. In addition to the two mentioned, there are the zeroth (0th) law (that heat won't flow between two items at the same temperature) and the third law (that you can never reach absolute zero).

1877 Year in which Austrian physicist Ludwig Boltzmann gave entropy a statistical formulation.

What Else?

What if Maxwell had a demon? *See page 124*

What if Einstein had invented a refrigerator? *See page 146*

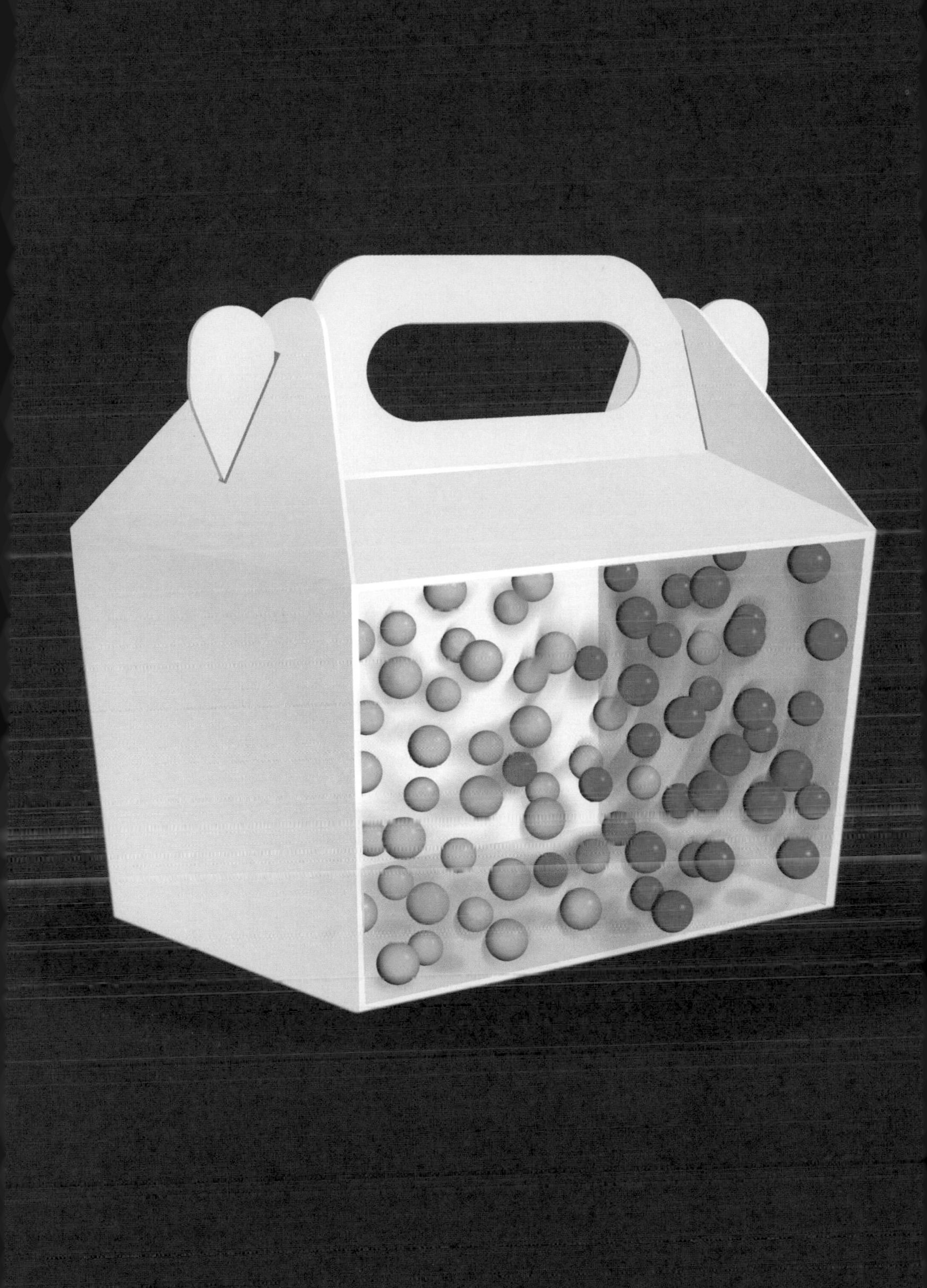

WHAT IF THERE WERE A MAXIMUM TEMPERATURE?

Brian Clegg

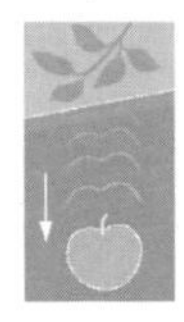

Temperature seems to be an everyday kind of measurement. But underlying temperature is a sophisticated concept that gets very interesting at the extremes. Temperature is often represented as a measure of the speed of the atoms or molecules in a substance. If you think of air, the gas molecules that make it up are constantly in motion, zipping from place to place. (Some move much faster than others – but temperature is a statistical measure, taking a picture across the whole set of molecules.) As the substance gets cooler, the particles become slower and slower. At the extreme we reach absolute zero (–273.15°C or –459.67°F), which is 0 K (Kelvin) in the scale that starts at this ultimate chill. If absolute zero could be reached, every atom or molecule would be in its least energetic state. In practice this can't be achieved. It is always possible to get a little closer to absolute zero, but quantum limitations prevent it being reached. What we hear less about is the other end of the temperature spectrum. Taking the simplistic idea that temperature is a measure of the velocity of particles in a material, there should be a maximum temperature. This is because special relativity says that the universe has an ultimate speed limit. Nothing can travel faster than the speed of light (around 300,000 km/sec [186,000 miles/sec]) in a vacuum. It is impossible for the component particles of a substance to move any quicker than this, however much energy you pour into the substance. This would seem to set an upper limit on temperature. However, temperature has a trick up its sleeve. It is really dependent on the kinetic energy of the atoms and molecules, and kinetic energy comprises both velocity and mass. Special relativity tells us that as an object moves faster, its mass increases, tending towards an infinite mass at the speed of light. So despite the limitation on the speed of particles, the temperature can still soar to infinity.

What Then?

Temperature is also influenced by the distribution of kinetic energy across particles. If all particles have the same energy, they have low entropy (the measure of the disorder in a substance) and that means low temperature. The maximum entropy for a body is when the particles are evenly spread across the possible energies. As the temperature reached infinite more and more particles would have similar high energies, the entropy would start to decrease and the temperature would flip to minus infinity.

What Gives?

30,000 K Top temperature of a lightning bolt – one of the hottest things we regularly experience on Earth. (Equivalents are 29, 725ºC/53,500ºF.)

15,000,000 K Temperature at the centre of the Sun. By comparison, the Sun's surface temperature is around 5,000 K. (Equivalents are 14,999,000ºC/26,999,000ºF.)

What Else?

What if you went back to the future? *See page 38*

What if something were colder than absolute zero? *See page 118*

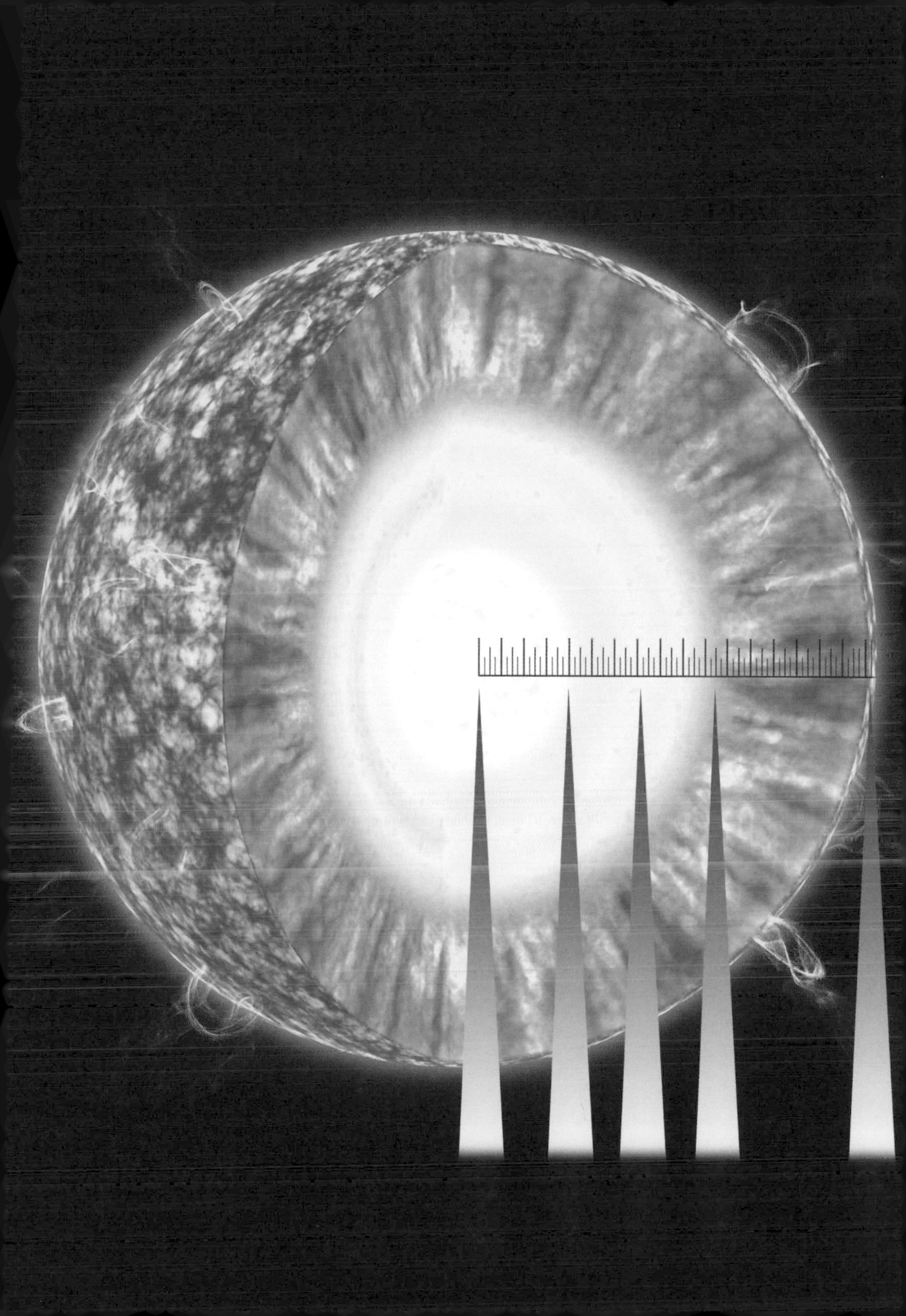

WHAT IF MAXWELL HAD A DEMON?

Simon Flynn

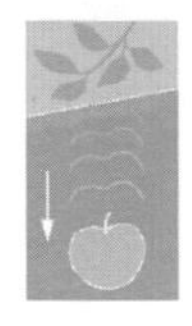

Heat engines convert energy in the form of heat into mechanical work. An example is the engine in your car. Unfortunately, the second law of thermodynamics ,'Heat energy always flows spontaneously from a hot to a cold system and never naturally the other way round', informs us that these engines can never be perfect because some of the energy will be lost to the surroundings – your car engine gets hot. The second law also makes it clear that for energy to flow from cold to hot, which is what happens in a refrigerator, work needs to be done. Albert Einstein and English astrophysicist Arthur Eddington were convinced the law could never be violated. However, Scottish theoretical physicist James Clerk Maxwell came up with a thought experiment designed to do just that. He had earlier shown that the molecules in a container of gas at any given temperature, and in thermal equilibrium, moved at different velocities and therefore had different energies – an example of kinetic (movement) energy. Temperature is related to the average of the sum of these energies: the higher the average energy, the higher the temperature. Maxwell proposed dividing the container in two (A and B) with a trapdoor between so that sides A and B were equivalent – they had the same temperature. The opening/closing of the door was to be controlled by a being, later known as Maxwell's demon, capable of perceiving each gas molecule. Over time, the demon would allow faster molecules (those with greater energy) to pass from A to B and slower ones (those with lower energy) from B to A. A temperature difference would be produced because the average energy of side B molecules would be greater than that of side A ones. This creation of a temperature difference would be in contradiction of the second law of thermodynamics (providing the demon didn't expend more energy than could be subsequently got out – for example, the trapdoor would have to be frictionless) and could be used to run a heat engine.

What Then?

In 1929, Hungarian-born physicist Leó Szilárd made a connection between the information processed by the demon and the effective energy produced by its actions. Over the next 80 years, scientists came to the consensus that the energy expended by the demon in determining the molecules' velocities would be greater than that created by the resulting temperature difference. In 2010 Japanese researchers reported an experiment in which a particle's energy was raised using information only, raising the possibility of an 'information-to-heat-engine'.

What Gives?

1843 Year in which the word 'thermodynamics' was introduced by William Thomson, later Lord Kelvin. The first law of thermodynamics can be stated as 'Energy can be neither created nor destroyed, only interconverted between forms.'

What Else?

What if the universe were random? *See page 20*

What if Einstein had invented a refrigerator? *See page 146*

What if motion were perpetual? *See page 150*

HISTORICAL

WHAT IF THE EARTH IS NOT FLAT?

The idea of a flat Earth met its Waterloo during the time of the ancient Greeks. Inspired by seeing that the Earth's shadow on the Moon was always circular, the Pythagoreans (the followers of the 6th/5th-century BC philosopher-mathematician Pythagoras of Samos) argued that the Earth was spherical. This chimed well with their view of beauty and harmony – they considered a sphere to be the perfect geometrical shape. In the 4th century BC Greek philosopher Aristotle extended, and in a sense settled, the argument through a series of further proofs that included an examination of the gradual disappearance of ships over the horizon and the observation that different stars can be seen as one moves north or south.

This begged the question, how big is the Earth? In the 3rd/2nd century BC Greek mathematician Eratosthenes of Cyrene was able to calculate a remarkably accurate figure for the Earth's circumference. He was aware that at noon on the day of the summer solstice a rod planted in the ground at Syene (modern-day Aswan, in Egypt) cast no shadow. This meant that the Sun was directly overhead. At exactly the same time, a rod at Alexandria cast a shadow showing the angle between the rod and the direction of the Sun's rays to be 1/50th of a circle. Using geometry, one can see that this angle is equal to the angle between the two rods from the Earth's centre, or 7.2 (1/50th of 360).

Therefore the distance from Syene to Alexandria, which Eratosthenes estimated at 5,000 stades, was 1/50th of the Earth's circumference. Depending on what length one judges a stade (one lap of a stadium) to be, Eratosthenes' circumference is no more than 16 per cent out – and possibly even 99 per cent accurate. Given the tools available at the time, this was an extraordinary achievement. Interestingly, Eratosthenes' method makes one particular assumption – the Sun must be far enough away from the Earth for the Sun's rays to be parallel. The ancient Greek view of the solar system was vital for this.

The Pythagoreans had a huge influence on 5th/4th-century BC Greek philosopher Plato, particularly regarding the power of geometry and mathematics to explain the world around us. There was a problem, however. The stars appeared to move in perfect circles, but the five planets then known (Mercury, Venus, Mars, Jupiter and Saturn) seemed to move in a non-uniform manner. Tellingly, the word 'planet' comes from the Greek word for 'wanderer'. This troubled Plato. He suggested that the Sun, Moon and planets moved in a combination of two circular movements, but this proved an unsatisfactory answer. One of Plato's students, the astronomer and mathematician Eudoxus of Cnidus, came up with the idea of extending the number of circular movements of each planet to four. It resolved many of Plato's issues, including providing a good approximation of a planet's retrograde motion when it seems to turn back on itself.

Aristotle appropriated Eudoxus' system and refined it. A fundamental change, however, was that he argued it was more than a mathematical argument but was tangibly real, that the spheres described existed and carried the planets around. Furthermore, his theory on the natural place of something (the effect we know to be gravity) demonstrated the Earth had to be at the centre of things.

The idea of a flat Earth was essentially dead, but at the heart of the ancient Greek solar system that replaced it lay two erroneous beliefs: that the Earth was at its centre and that the planets moved in a circular motion. It wasn't until 1543 that Polish astronomer Nicolaus Copernicus struck the crucial blow on the first of these by declaring all planets to orbit the Sun. The second was resolved in the 17th century when German mathematician and astronomer Johannes Kepler formulated his three laws of planetary motion and showed the orbits of the planets to be elliptical in shape.

WHAT IF A MIRROR DID NOT REVERSE LEFT AND RIGHT?

Brian Clegg

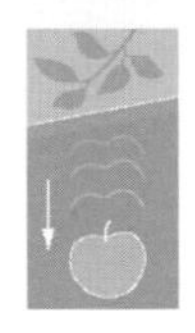

One of the first aspects of classical physics to be developed was basic optics. It took a while to be clear where light originated – the ancient Greeks thought it came from the eye, rather than a light source such as the Sun – but from surprisingly early on a neat picture emerged: rays of light hit an object, bounced off it like a ball bouncing off a wall and then travelled in a straight line from the object to the eye. This was verified repeatedly and became the basic law of optics. But there was a problem. The reflection process was symmetrical on a basic reflector like a flat mirror. It didn't matter which direction a light ray hit such a mirror, it would reflect off in the same direction at the same angle with which it arrived. A ray heading left to right as you look at the mirror would reflect off left to right. A ray heading top to bottom would continue on its top-to-bottom path – though in both cases, they would go from heading towards the mirror to heading away. And yet there was a clear contradiction in what people actually saw: in a flat mirror, the symmetry was broken. The mirror reversed things left and right, turning a left glove into a right, and swapping the side a car's driver sat – yet there was no equivalent reversal top and bottom. What made them different? When faced with the challenge of explaining this, many of us first think it must be linked to the asymmetrical left/right orientation of our eyes. Yet this idea falls down as soon as you look at a mirror sideways on and don't see top and bottom swapped. In fact it is just our perception that images are reversed left and right. They are actually turned inside out.

What Then?

We imagine left/right reversal because we think of our body being rotated through 180 degrees to become our mirror self – but this is not what the optical system of the mirror does. It's as if it turns a rubber mould of your face inside out. The tip of your nose, the closest bit of you pointing into the mirror, flips inside out to become your reflection's nose, its closest part pointing out of the mirror.

What Gives?

Quantum reflections In classical physics light bounces on to and off a mirror at the same angle. Quantum physics shows that it reflects off at every possible angle, but usually the odd directions cancel out leaving the classical reflection.

Front or back? Hold up a book in front of a mirror – show the mirror its front cover and you will see that in the reflection the spine is on the right, meaning it is the reflected book's back cover.

What Else?

What if light is not a wave? *See page 26*

What if there were not seven colours in a rainbow? *See page 130*

WHAT IF THERE WERE NOT SEVEN COLOURS IN A RAINBOW?

Simon Flynn

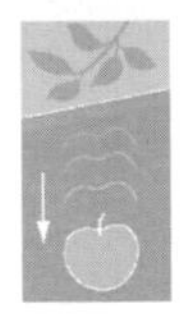

In *Meteorologica*, the Greek philosopher Aristotle argued that rainbows had just three colours – red, green and violet. Around 2,000 years later, English physicist Isaac Newton conducted his first experiment separating white light using a prism. He initially wrote of five primary colours: red, yellow, green, blue and violet. He later added orange and indigo, in part because he viewed colour as being like music, capable of similar harmonies, the seven colours matching the seven notes of a Dorian scale (an ancient Greek musical scale). At the time, this connection made Newton's theory of optics easier for people to accept (curiously, in *De sensu*, Aristotle made similar reference to musical harmonies in discussing colour). But the description of the seven colours in a rainbow helped establish a common misconception – it was quickly forgotten that Newton had also talked of an 'indefinite variety of intermediate gradations' between the primary colours. We now know visible light to be a continuous spectrum, that it's an uninterrupted band of colours whose wavelengths range from about 400 nm to 700 nm. This is just a tiny part of the electromagnetic spectrum, which also includes radio waves, microwaves and gamma rays. These are all electromagnetic waves travelling at the speed of light, but with differing frequencies and wavelengths. Visible light is merely the range of radiant energies to which human eyes are sensitive. Penguins and honeybees, for example, are able to see into the ultraviolet range of light and rattlesnakes and bed bugs can see into infrared. If we could see radio waves, the world would look very different. It turns out we're far more sensitive to light in the middle of the range, roughly from light blue to orange. This still raises the question of how many colours we're able to distinguish. One estimate is about 100. It's certainly a lot more than seven.

What Then?

The universe is actually aglow if viewed in the microwave region of the electromagnetic spectrum. This cosmic microwave background radiation (CMBR), as it is known, is believed to be a legacy of the big bang – and is a key piece of evidence in support of this theory. As the universe expanded, the initial high-energy radiation stretched and cooled. CMBR is responsible for the universe having a residual temperature of 2.7K. (2.7K is -270.45°C or -454.81°F.)

What Gives?

Fruity Orange didn't exist as a specific colour until the 15th century and takes its name from the fruit.

299,792,458 m/sec Speed of light. All electromagnetic radiation travels at this speed. (299,792,458 m/sec is 983,571,056 ft/sec.)

What Else?

What if we could see the atom? *See page 68*

What if we could replace electrons with light? *See page 70*

What if a mirror did not reverse left and right? *See page 128*

WHAT IF WATER BOILED AT –70°C?

Simon Flynn

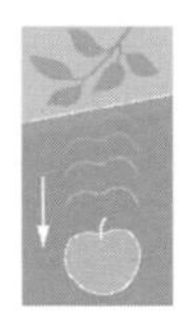

As we know, the boiling point for water, H_2O, is 100°C (212°F). Oxygen is the first element in group 16 of the periodic table. As we descend the group, the boiling points of each element bonded to two hydrogen atoms runs as follows: H_2S (hydrogen sulfide) is –60°C (–76°F), H_2Se (hydrogen selenide) is –50°C (–58°F) and H_2Te (hydrogen telluride) is –2°C (–28°F). There seems to be a pattern here. If we didn't know the boiling point of water and used these to make a prediction as to what it would be, then we'd expect the answer to be about –70°C (–94°F). If it was, life as we know it simply could not exist. The reason for the huge discrepancy is something called hydrogen bonding. An oxygen atom is said to be very electronegative. (Electronegativity is the power of an atom in a molecule to attract electrons.) When the oxygen atom is bonded to hydrogen, which is much less electronegative, the electrons in the bond are attracted more towards the oxygen atom. This results in what is known as a dipole, in which each hydrogen atom possesses a partial positive charge and the oxygen atom has a partial negative charge. It could be thought of as being like a very weak magnet. Bring together a whole collection of water molecules, each with partially positive hydrogen atoms and partially positive oxygen atoms, and the hydrogen atoms will make a link with the oxygen atoms between molecules. This force between molecules results in a higher melting and boiling point for H_2O than would be expected from looking at the boiling points of H_2S, H_2Se and H_2Te. (S [sulphur], Se [selenium] and Te [tellurium] are much less electronegative than O [oxygen], so there is little inducement of dipoles in those molecules.) You get the higher melting and boiling points because more energy has to be put in to disrupt the bonds between molecules as well as the main ones within the H_2O molecules between the oxygen atom and the two hydrogen atoms.

What Then?

Not only is water vital to life, hydrogen bonds in general are, too. DNA consists of two giant molecules held together by hydrogen bonding. Hydrogen bonds are also partly responsible for large proteins, such as insulin, folding into the shape they need to have in order to work. And the high-strength material Kevlar® owes much of its desired property to the hydrogen bonding between its polymer chains.

What Gives?

1/20th Strength of a hydrogen bond compared to that between hydrogen and oxygen in a water molecule itself.

NH_3 Hydrogen bonding can also occur when hydrogen molecules are bonded to nitrogen or fluorine (for example, NH_3 and HF).

What Else?

What if we had GUTs and TOEs? *See page 58*

What if something were colder than absolute zero? *See page 118*

2

TECHNO

TECHNOLOGY

INTRODUCTION
TECHNOLOGY

Technology extends our basic human capabilities, enabling us to do things that are impossible with our hands and bodies alone. It may seem that technology is out of place in a book like this – yet most technology is driven by physics, and it is through technology that most of us directly experience the results of physical research in laboratories. We might never visit CERN, but we can see physics in action all around us.

Think, for instance, of transport. Whenever we get an aeroplane or a car moving we are relying on the laws of motion established by English physicist Isaac Newton in the 17th century. An aircraft, for instance, uses Newton's third law ('every action has an equal and opposite reaction') to get moving. Its engines push on the air and the air exerts an equal and opposite push on the engines (hence on the aeroplane), propelling it forwards. Every time we push down the accelerator in the car, we are making use of thermodynamics in the conversion of heat in the engine to work and the relationship between energy and the mass of the car that it accelerates.

Sometimes the role of physics is hidden away until we think about how a piece of technology works. A refrigerator is an excellent example. It is easy enough to heat things up, but it isn't exactly obvious how to make the temperature of something fall until you understand both thermodynamics and what happens to matter when substances expand.

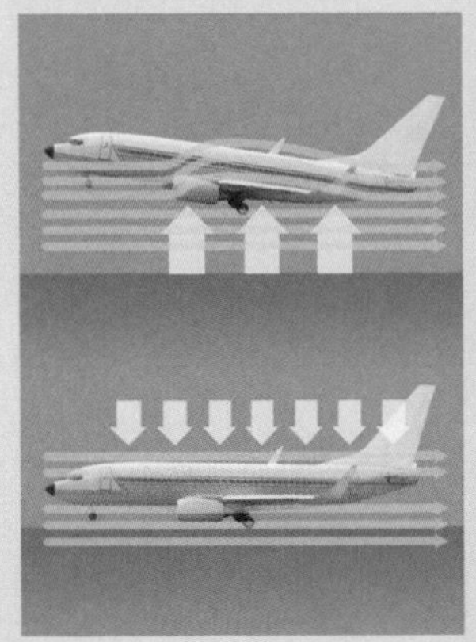

Although we are surrounded by technology based on traditional physics – machines like the wheel, the lever and the screw will always be important – our modern lives also rely on technology that makes use of the latest physics. Electronics, for example, is based on quantum theory. Even the earliest forms of electronics like a vacuum tube (valve) had to control the flow of electrons – quantum particles. Modern solid-state electronic devices have an even tighter link to the intricacies of the quantum world.

Then there are robots. Some are simple but dextrous, like the robots used to assemble cars or run automated warehouses. Others attempt to be more like a thinking creature. And it is in their future that technology takes on one of the biggest challenges for physics – identifying the fundamental physical principle behind consciousness and how the mind works. In trying to improve robotic technology we also have the chance to get a better understanding of our own brains.

Many robots are large, but technology is also increasingly exploring the other end of the size spectrum. Nanotechnology exploits the potential for everything from simple materials to complex devices on a scale similar to a virus. Here the physical laws work very differently to our large-scale world. Even sheets of carbon just an atom thick have the potential to deliver huge benefits – all technology with physics at its heart.

WHAT IF ROBOTS WERE CONSCIOUS?

Angela Saini

In the mysterious hinterland between science and philosophy, few things are as difficult to understand as human consciousness. Many engineers trying to synthetically reproduce consciousness believe it to be linked to intelligence. But those in the field of artificial intelligence (attempting to build machines that can carry out tasks that usually require the power of the human mind) have yet to define what intelligence is. Many branches of research are looking at the problem of consciousness and intelligence from different angles. For some, the key to artificial intelligence lies in getting computers to solve logical problems using approaches such as Bayesian reasoning, which uses best-guess probabilities to predict an outcome, the way humans seem naturally to work through decision-making in our heads. Others believe that the secret lies in building robots that can sense the world around them as people do, using artificial neural networks that mimic the biological links in the brain. Which, if any, approach is successful, all engineers agree that it will take huge amounts of calculating power. The biggest, fastest computers of the age – the supercomputers – have already achieved remarkable feats. In 1997 one called Deep Blue, built by American technology giant IBM, beat world chess champion Garry Kasparov in a chess match (it had lost its first match against him in 1996). In 2011, Watson, another IBM supercomputer, competed on the American game show *Jeopardy!* and won US$1 million. Today language is thought to be vital to artificial intelligence, which is why many researchers are developing systems to help machines understand human vocabulary and grammar. One of the latest artificial intelligence endeavours has been to design a machine that can understand stories. Professor Patrick H. Winston at MIT in Cambridge, Massachusetts, has designed software called Genesis that is able to explain the plotlines behind Shakespeare's tragedy *Macbeth*.

What Then?

In 1950 British computer scientist Alan Turing suggested that one way of judging a computer's intelligence might be to study whether a person could tell the difference between it and a human being, if they were in conversation. Machines have already claimed to pass this Turing Test, but it isn't enough of a measure of true human-like intelligence. Before inventing robots that can exhibit consciousness, scientists have to figure out what intelligence really is.

What Gives?

1968 Year in which the artificially intelligent computer HAL 9000 starred in the Hollywood film *2001: A Space Odyssey*.

2011 Year in which Apple introduced its iPhone assistant, Siri, to the public. Siri is unnervingly like a helpful human, recognizing voice commands given by an iPhone user.

20,000 trillion Number of calculations per second performed by the world's most powerful supercomputer, Titan.

What Else?

What if we could compute with quanta? *See page 30*

What if we could replace electrons with light? *See page 70*

WHAT IF WE WERE MENACED BY GREY GOO?

Angela Saini

A fairly young branch of science, nanotechnology is all about working with materials on the tiniest possible scale: that of atoms and molecules, which can be as small as one-billionth of a metre across. The unsettling phrase 'grey goo' is commonly credited to Prince Charles, who in 2003 stoked alarm about the unknown prospects of nanotechnology. But the words actually originated in the work of one of the world's first nanotechnologists: in his book *Engines of Creation* (1986), American engineer Eric Drexler vividly imagined the possibility of atomic-scale machines sophisticated enough to copy themselves independently – just as living creatures do. These tiny self-replicators might multiply exponentially, he suggested, sucking in the raw materials they needed for their growth from the natural environment around them. Drexler added that, to our naked eyes, these machines would be so small that they would look like nothing more than a growing pile of dust – although not necessarily grey or gooey. Best-selling American author Michael Crichton took Drexler's vision one nightmarish step further in his novel *Prey* (2002), in which a cloud of microscopic predators turn on the same scientists who created them. In reality, however, this dark vision of runaway miniature robots is as distant today as it was in 1986. Today, nanoparticles are widely used in medicine, cosmetics and clothing. Because of their tiny size, the body finds nano-sized drugs easier to absorb, for instance; antibacterial nanosilver helps sportswear smell sweeter for longer. Very few mechanical nanomachines have been developed, and even these have been used mainly in biomedical devices to fight disease. Drexler's famous phrase has become a fading spectre over nanotechnology, with the iconic engineer himself even reportedly wishing that he had never used it. Nanotechnology, it seems, has turned out to be rather more mundane than menacing.

What Then?

Iconic engineer Eric Drexler wrote in 2004 that 'deliberate human use of powerful systems can lead to serious trouble' – like Goethe's multiplying brooms in his poem 'The Sorcerer's Apprentice'. Runaway self-replicating nanomachines cannot be invented by accident, according to Drexler, but there is a (distant, theoretical) chance they might be developed deliberately. So the onus is on scientists and governments to strictly regulate nanotechnological advances to ensure this never happens.

What Gives?

1959 Year in which American physicist Richard Feynman first proposed the idea of nanometre-scale machines.

Just over half The people in the UK surveyed in 2008 who found nanotechnology morally acceptable. Some 30 per cent of a sample of the US population agreed with the proposition.

Middle Ages Period in which nanoscale particles of gold and silver were used in stained glass windows.

What Else?

What if we could see the atom?
See page 68

What if robots were conscious?
See page 138

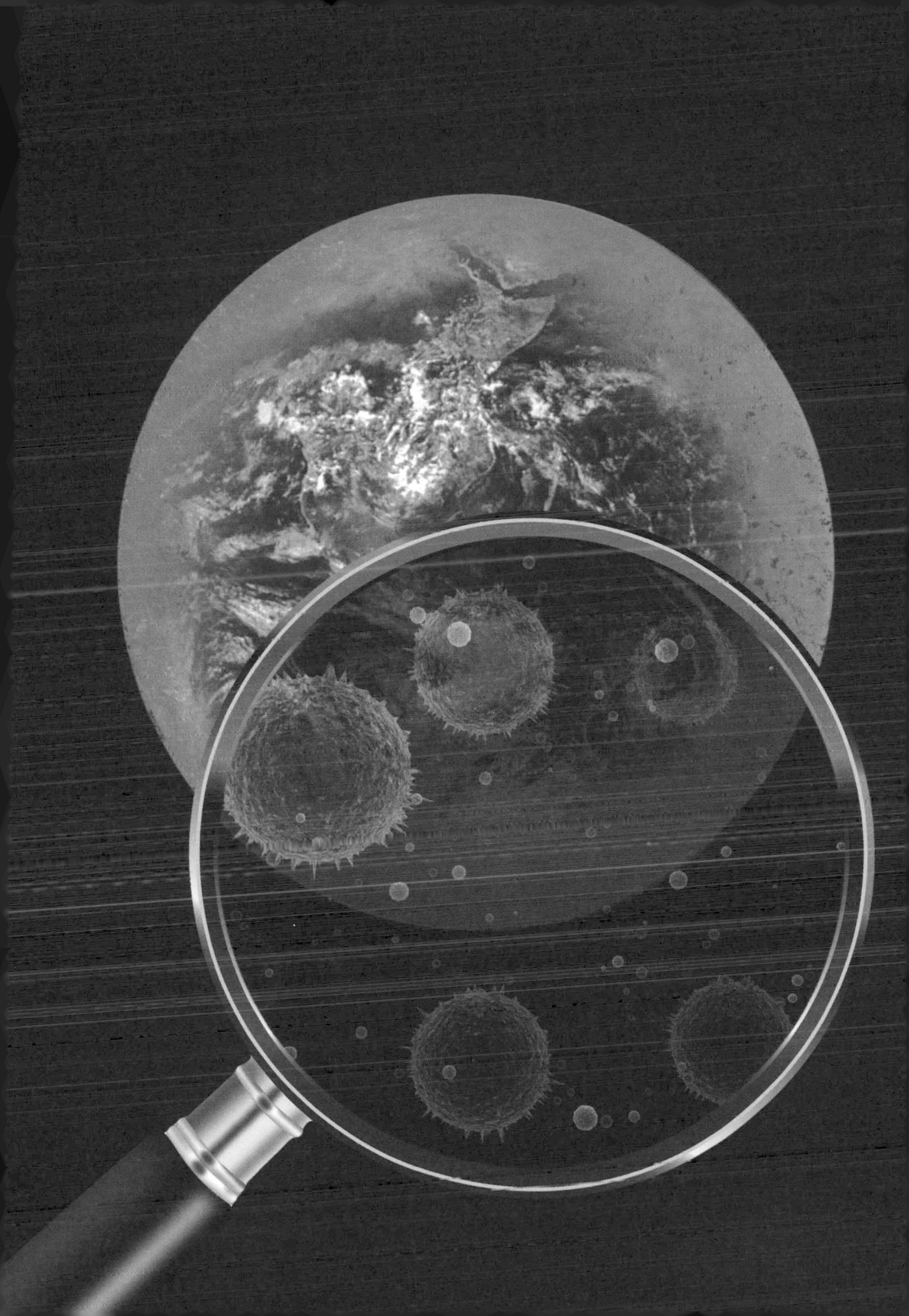

WHAT IF CARBON COULD CHANGE THE WORLD?

Angela Saini

Once in a while, Nature offers up a material so powerful in its technological applications that you'd think it had been dreamed up in a lab. Graphene is one of these miracle materials. Made from geometric, atom-thin sheets of carbon, it's derived from naturally occurring graphite, popularly known for its use in pencils. Graphene's array of amazing properties include super strength (it is 200 times stronger than structural steel, according to researchers at Columbia University in New York) and super lightness (a square-metre sheet of graphene would weigh less than one-thousandth of a gram and be practically transparent). Since its discovery in 2004 by Andre Geim and Konstantin Novoselov – Russian-born researchers at Manchester University, England – graphene has been proposed for thousands of potential uses. Tennis rackets made using graphene are lighter, for example. Graphene in vehicle tyres could make them stronger. But what has excited people most about graphene is that it conducts electricity better than any metal, therefore it might be a stronger, lighter, more flexible alternative to silicon in microelectronics. In 2007 Geim and Novoselov suggested that plastics containing just 1 per cent graphene by volume would be able to conduct electricity. Since then, the American technology company IBM has developed high-speed circuits made out of graphene. But before you start imagining what this new generation of strong and bendy gadgets might look like, remember that graphene is a relatively new discovery, so its promises are yet to be proven. One of the hitches already found with its use in electronics is that in graphene, unlike silicon, the flow of electricity can't be switched off. However, South Korean company Samsung claims to have developed a device that can switch off the current in graphene. Large amounts of money are being pumped into graphene research all over the world – this miracle material could prove to be the real deal.

What Then?

One of the discoverers of graphene, Andre Geim, suggested that in the future the new material could be used in ways similar to the way plastics are employed today. But that potential has yet to be realized. Much of the work on graphene has been done on a tiny scale. Graphene is very expensive to make in large quantities, and it's possible that its amazing properties may not translate so well on a larger scale in the real world.

What Gives?

2010 The year Russian-British physicists Andre Geim and Konstantin Novoselov won the Nobel Prize in Physics for discovering graphene.

20 per cent The amount by which a sheet of graphene can extend when stretched.

Better than a diamond Graphene conducts heat better than any other material, including diamond.

What Else?

What if we could see the atom? *See page 68*

What if we could replace electrons with light? *See page 70*

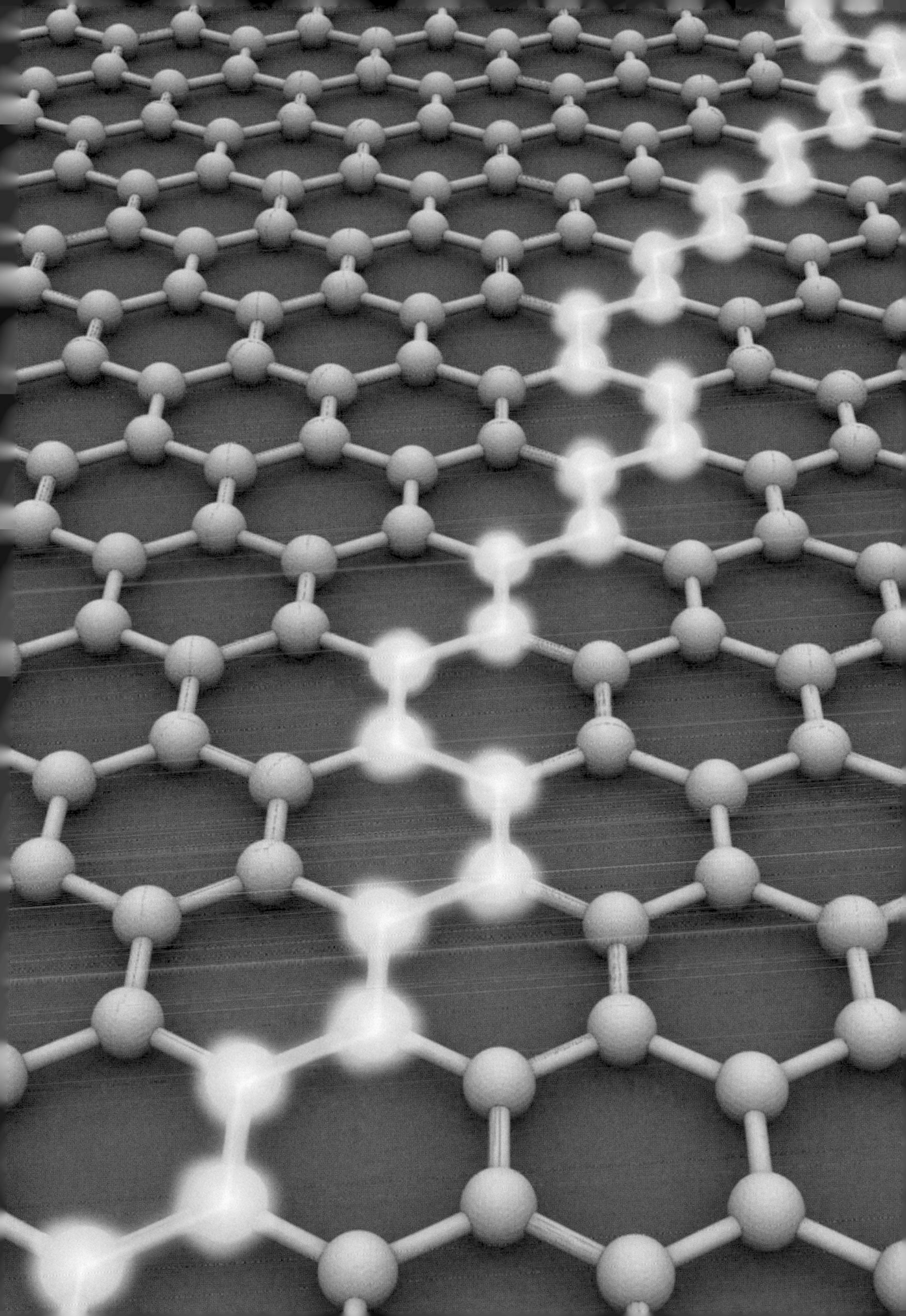

WHAT IF AIRCRAFT WINGS DID NOT WORK?

Brian Clegg

The jet engine that gives an airliner its thrust is only half the story of getting the aeroplane into the air. The other essential piece of technology is the wing, which provides lift. There is a well-known explanation of how these work called the Bernoulli effect, which is very elegant but has one significant problem – it's wrong. The Bernoulli effect explanation goes something like this. Wings are specially shaped so air travelling over the top has further to go than air passing under the wing. This means the air over the top has to go faster to keep up. In the process the air over the top thins out, which means it exerts less pressure on the wing than the air at the bottom. With lower pressure above than below, the wing is pushed up. Outcome: lift. The problem with this argument is that there is no reason why air travelling over the top of a wing should try to 'keep up' with air round the bottom – and in practice it typically flows quicker over the top than it would need to catch up, anyway. There is a lift from the Bernoulli effect – it is real – but it is nowhere enough to keep a 400-tonne aircraft flying. The main lift from a wing derives from the same effect that jet engines use: Newton's third law of motion (as defined by English physicist Isaac Newton in the 17th century), which says that every action has an equal and opposite reaction. In the engine, air is blasted out of the back of the jet. As a result of the push back on the air, the engine (and hence the aeroplane) experiences an equal thrust forwards. As for the wing, its shape and angle is designed so that the air it moves through is pushed down. If the air is pushed down, the wing is pushed up. We have lift.

What Then?

Most of what happens to a plane when it is in the air involves five key forces. Gravity is always pulling it downwards; to counter this, when the plane is moving forwards, lift pushes up on the wings. Thrust from the engines pushes the plane forwards, and drag pulls the plane back because of air resistance. The joker in the pack is turbulence, which pushes the plane all over the place, typically when crossing areas of different temperature or air-movement speeds.

What Gives?

64 metres Wingspan of a 747 airliner – this figure (211ft) is around twice the distance flown by the Wright Flyer, the first powered aeroplane, on its maiden flight in 1903.

Invisible vortexes
Effect of wing tips passing through air – like water running down a plughole. Aeroplanes have to wait until after the previous plane has finished using a runway because wing tips create invisible vortexes in the air that cause turbulence.

What Else?

What if you went back to the future? *See page 38*

What if the Earth is not flat? *See page 126*

WHAT IF EINSTEIN HAD INVENTED A REFRIGERATOR?

Brian Clegg

At first sight, a refrigerator seems to defy the second law of thermodynamics, which states that heat flows from a hotter to a colder body. In a fridge, heat is extracted from the cold space within the device and radiated to the warmer outside. This is possible because the second law is limited to closed systems – ones where no energy flows in or out. But a fridge will only function if it is plugged in. It needs energy from outside to work. Fridges usually depend on the expansion of a substance called a refrigerant, a compound such as tetrafluoroethane. This starts off as a vapour, which is compressed and allowed to cool by exposure to the air outside the refrigerator, giving off heat. As it cools, the refrigerant liquefies. It is then forced through a small aperture into a low-pressure chamber where part of it flash-evaporates. This takes energy out of the remains of the refrigerant, which is still liquid, dramatically cooling it. Air that passes over pipes with the cooled refrigerant running through them is taken into the interior of the fridge, chilling the air. This is the mechanism that operates in most refrigerators today – but in the early days the refrigerant was toxic and in the 1920s a family in Berlin, Germany, was killed when a seal broke on their fridge. This inspired the unlikely pairing of Albert Einstein with Hungarian-born physicist Leó Szilárd to invent a totally different mechanism for refrigeration that had no moving parts and that operated at constant pressure, rather than requiring the high compression of a traditional fridge. It may seem strange that Einstein, the ultimate theorist, was involved in such an invention, but his first job had been in the Swiss Patent Office, checking applications to see if they made sense. It seems likely that this experience made him the perfect partner for his former student Szilárd, best known for coming up with the idea of the nuclear chain reaction.

What Then?

Although Einstein and Szilárd's invention received many patents, it was not widely used, but is still being considered for future refrigerators in regions where electricity supplies are limited or unavailable, as the Einstein refrigerator only requires a source of heat to run, rather than electricity, so it can be powered by anything from gas to solar energy. Instead of relying on a compressor it uses a mix of two compounds, one of which can be extracted from the mix to suddenly drop the pressure.

What Gives?

1926 The year in which Einstein and Szilárd invented their absorption refrigerator.

1781541 The US patent number of the Einstein refrigerator, filed in 1927 and issued in 1930. It continues to be referenced to the present.

What Else?

What if something were colder than absolute zero? *See page 118*

What if there were such a thing as a free lunch? *See page 120*

HISTORICAL
WHAT IF ELECTRICITY AND MAGNETISM ARE NOT SEPARATE?

While electricity and electrical gadgets seem to colonize ever more numerous areas of our lives, our daily experience of magnetism rarely extends beyond fridge magnets. But although they look so different at face value, electricity and magnetism are intimately connected: you can't have electricity without magnetism or magnetism without electricity. An electric current flowing through a wire creates a magnetic field around it, which is the basis for how motors work, and if you move a magnet near to a wire you can induce an electric current to flow in the wire – hey presto, you are generating electricity.

The relationship between electricity and magnetism was first discovered, almost by accident, by Dutch professor of physics Hans Christian Ørsted in 1820. While he was giving a lecture to demonstrate how an electric current heats up a wire, he noticed that a nearby compass needle was deflected by the wire when the current was switched on or off. He later investigated it and published the discovery, stirring up plenty of interest in the scientific community.

The work inspired French physicist André-Marie Ampère, who guessed that if a current in a wire could exert a force on a compass needle, making it move, two such wires should also interact magnetically. He showed that two parallel wires carrying currents in the same direction attract one another, whereas when the same wires carry currents running in opposite directions they repel. He went on to formulate Ampère's law, which describes the magnetic field generated by a current-carrying wire. The unit for quantifying electric currents is named after Ampère and its definition comes from exactly these experiments, in terms of the force two parallel wires exert on each other.

Experiments in the 1830s by English chemist and physicist Michael Faraday were also important. He discovered electromagnetic 'induction': how to induce an electric current to flow in a coil of wire by moving a magnet inside it. Key to this is the movement of the magnetic field – if you stop moving the magnet, the current in the wire disappears. This is exactly how we generate electricity in power stations. We burn coal or use the energy from falling water at a hydroelectric plant to drive magnets to move relative to coils of wire. Other experiments at this time also laid the foundations for building the modern electric motor, generator and transformer.

Following on from these experiments, Scottish theoretical physicist James Clerk Maxwell generalized the behaviour of electric and magnetic fields into a neat package of formulas, called Maxwell's equations, which work for any situation. They represent a towering intellectual achievement, on a par with the laws of motion formulated by English physicist Isaac Newton, and show how electricity and magnetism together make up one of the four fundamental forces in the universe, which we call the electromagnetic force. (The other three are gravity, the weak nuclear force and the strong nuclear force.)

The most amazing feature of Maxwell's equations is that they show how a time-varying field – of either kind, magnetic or electric – induces a field of the other kind in neighbouring regions of space. This predicted the existence of electromagnetic waves – time-varying electric and magnetic waves that don't require a medium to move, but can travel through free space. These waves make up the electromagnetic spectrum, and include light, radio waves, infrared and X rays, and are also fundamental to today's technology.

WHAT IF MOTION WERE PERPETUAL?

Simon Flynn

Integral to the idea of perpetual motion are English physicist Isaac Newton's three laws of motion. Newton's first law tells us that a body will stay in a state of rest or will continue to keep moving if already doing so unless an external force is applied. So, a golf ball on the ground remains stationary until you apply a force to it by hitting it with a club. As soon as you do hit it, other forces start to act against it – crucially, air resistance and gravity – and there is a limit to how far it will travel. One would expect it to travel a lot further if hit in space because of there being fewer and weaker forces subsequently acting against it. The first difficulty, then, with creating a perpetual motion machine on Earth are opposing forces such as friction. Yet even in space there will be some force, such as gravity, acting on a body – although the force is very, very weak. Further problems arise when the first two laws of thermodynamics are considered. The first law refers to the conservation of energy. In an isolated system, you can't get out more than you put in. The second law prohibits the possibility of all useful energy being converted into work – some will always be lost to the surroundings by heat. There are, however, many things in nature that appear to be undergoing perpetual motion – the Moon, for example. For billions of years, it has journeyed around the Earth at an apparently constant speed. But very detailed measurements show its motion to be changing. The question ultimately boils down to what is meant by 'perpetual'. If this is understood to mean forever, then science tells us perpetual motion is impossible, except at the level of quantum particles which can never truly stop moving. But if, instead, it refers to a very long time, then perpetual motion is almost certainly already happening. The key is making it work for us.

What Then?

Given the example of the Moon, perhaps the closest the human race has got to producing perpetual motion is the spacecraft *Pioneer 10*. Launched in 1972, it is now heading out into deep space, and though it has been overtaken by *Voyager 1*, has been in motion longest. Although contact was lost in 2003, scientists estimate that it will reach somewhere close to Aldebaran, a star more than 60 light years away, in about 2 million years.

What Gives?

12 km/sec *Pioneer 10*'s velocity with respect to the Sun. (12 km/sec is 7.5 miles/sec.)

4.5 billion years Estimated time that the Moon has been orbiting the Earth.

What Else?

What if there were such a thing as a free lunch? *See page 120*

What if Maxwell had a demon? *See page 124*

WHAT IF YOU HAD THE PHILOSOPHER'S STONE?

Angela Saini

History is studded with tales of a philosopher's stone, a legendary substance that could turn everyday metals such as iron and lead into silver or gold. Before the dawn of modern chemistry, the quest for this – known as alchemy – was treated like a genuine science, and helped to shape our understanding of the chemical elements. But once it seemed clear that alchemy might be physically impossible, the search for the philosopher's stone became little more than a pseudo-scientific dead end. Well, until recently. Today we know that each element is made of atoms with a certain weight and balance of particles within them that decides whether they are iron, for example, or gold. This is why it is so difficult to change one atom into another. It would mean breaking open the heart of an atom and removing or adding something. Difficult, that is, unless the element happens to be radioactive. Potassium, for instance, has a radioactive form that naturally turns into a type of argon by popping out tiny, subatomic particles. With some effort, scientists can force other elements to do the same. Indeed, while traditional alchemy may have been retired to the history books, turning lead into gold has already been done. American chemist Glenn Seaborg, a pioneer in nuclear physics who helped create nine new elements, achieved the dream in 1980. At the Lawrence Berkeley Laboratory in California he removed enough particles from inside a few thousand lead atoms to turn them into gold. Unfortunately, the process was too expensive to be worthwhile in the real world. More routinely, there is a form of uranium that, when attacked with subatomic particles, breaks apart into different – and some medically useful – elements. Effecting this is dangerous, expensive and tough. But in 2011 scientists at Osaka University, Japan, and at the Japan Synchrotron Radiation Research Institute found that they might be able to achieve the same end more easily by 'shaking' uranium atoms using laser beams.

What Then?

Today, one promising use of transmutation is to make safe the thousands of tonnes of dangerous radioactive waste produced by nuclear power plants. At the moment, this waste has to be carefully stored for hundreds of years, but engineers are working on ways to break down the waste elements – uranium, plutonium and some other highly radioactive by-products – into less deadly ones. One possibility is to use a particle accelerator, but this may turn out to be too costly.

What Gives?

1901 The year that scientists first described one radioactive element decaying into another as transmutation.

Beyond death
The legend of the philosopher's stone held that, in addition to transforming materials into silver or gold, it might also be an elixir for immortality.

What Else?

What if we had GUTs and TOEs? *See page 58*

What if everything were made of string? *See page 84*

NOTES ON CONTRIBUTORS

Jim Al-Khalili, OBE, is a British scientist, author and broadcaster. He is a professor of Physics at the University of Surrey where he also holds a chair in the Public Engagement in Science. He is president of the British Humanist Association. He is the author of the popular science books *Paradox: The Nine Greatest Enigmas in Physics* (Black Swan, 2013) and *Quantum: A Guide for the Perplexed* (Phoenix, 2012).

Brian Clegg read Natural Sciences, focusing on experimental physics, at Cambridge University. After developing hi-tech solutions for British Airways and working with creativity guru Edward de Bono, he formed a creative consultancy advising clients ranging from the BBC to the Met Office. He has written for *Nature*, *The Times*, and *The Wall Street Journal* and lectured at Oxford and Cambridge Universities and the Royal Institution. He is editor of the book review site www.popularscience.co.uk, and his own published titles include *A Brief History of Infinity* (Robinson Publishing, 2003) and *How to Build a Time Machine* (St. Martin's Griffin, 2013).

Frank Close, OBE, is Professor of Physics at Oxford University and Fellow of Exeter College, Oxford. He was formerly Head of Theoretical Physics Division, at Rutherford Appleton Laboratory, and Head of Communications and Public Education at CERN. His research is into the quark and gluon structure of nuclear particles, where he has published over 200 papers in the peer-reviewed literature. He is a Fellow of the American Physical Society, and of the British Institute of Physics, and won the society's Kelvin Medal in 1996 for his outstanding contributions to the public understanding of physics. He is the author of many books including *Antimatter* (Oxford University Press, 2010), *Neutrino* (Oxford University Press, 2012)– short-listed for the Galileo Prize in 2013 – *Nothing* (Oxford University Press, 2009), the best selling *Lucifers Legacy* (Oxford University Press, 2000) and most recent *The Infinity Puzzle* (Oxford University Press, 2011).

Rhodri Evans studies and researches in extra-galactic astronomy. Rhodri has, for over 16 years, been involved in airborne astronomy, and is a key part of the team building the facility far-infrared camera for SOFIA. He is also involved in research into star-formation and cosmology, and is a regular contributor to television, radio and public lectures. Rhodri runs the blog www.thecuriousastronomer.wordpress.com.

Simon Flynn is a science teacher and author of *The Science Magpie: A Hoard of Fascinating Facts, Stories, Poems, Diagrams and Jokes Plucked from Science and its History* (Icon, 2012).

Sophie Hebden is a freelance science writer based in Mansfield, UK. She combines writing about physics with looking after two small children. She has written for *New Scientist* and the *Foundational Questions Institute*, and is former news editor for *SciDev.Net*. She holds a PhD in space plasma physics, and a masters in science communication.

Angela Saini is an independent science journalist based in London, and the author of *Geek Nation: How Indian Science is Taking Over the World* (Hodder & Stoughton, 2011). Her writing has appeared in *New Scientist*, *Wired* and the *Guardian*, and she regularly presents science shows on BBC radio. Angela has won a number of awards for her journalism, including Best News Story in 2012 from the Association of British Science Writers. She has a Masters in Engineering from Oxford University and is a former Knight Science Journalism Fellow at the Massachusetts Institute of Technology.

WHAT IF RESOURCES

Books

Before the Big Bang
Brian Clegg
(Saint Martins Griffin, 2011)

Black Holes and Time Warps: Einstein's Outrageous Legacy
Kip S. Thorne
(W. W. Norton, 1994)

Black Holes, Wormholes and Time Machines
Jim Al-Khalili
(Taylor & Francis, 2012)

A Brief History of Infinity: The Quest to Think the Unthinkable
Brian Clegg
(Robinson Publishing, 2003)

A Brief History of Time
Stephen Hawking
(Bantam, 2011)

Build Your Own Time Machine: The Real Science of Time Travel
Brian Clegg
(Gerald Duckworth & Co., 2013)

Compendium of Theoretical Physics
Armin Wachter and
Henning Hoeber
(Springer, 2005)

Dice World: Science and Life in a Random Universe
Brian Clegg
(Icon Books, 2013)

The Elegant Universe: Superstrings, Hidden Dimensions and the Quest for the Ultimate Theory
Brian Greene
(Vintage, 2000)

The Fifth Essence
Lawrence Krauss
(Vintage, 1990)

The Infinity Puzzle
Frank Close
(Oxford University Press, 2011)

In Search of Schrodinger's Cat
John Gribbin
(Black Swan, 1985)

Introducing Infinity
Brian Clegg
(Icon Books, 2012)

The God Effect: Quantum Entanglement, Science's Strangest Phenomenon
Brian Clegg
(Saint Martins Griffin, 2009)

The New Cosmic Onion: Quarks and the Nature of the Universe
Frank Close
(Taylor & Francis, 2006)

Paradox: The Nine Greatest Enigmas in Physics
Jim Al-Khalili
(Black Swan, 2013)

Particle Physics: An Introduction
Frank Close
(Oxford University Press, 2004)

Quantum: A Guide for the Perplexed
Jim Al-Khalili
(Phoenix, 2012)

The Quantum Universe: Everything That Can Happen Does Happen
Brian Cox and Jeff Forshaw
(Penguin, 2012)

The Road to Reality
Roger Penrose
(Vintage, 2005)

Why Does $E=MC^2$?
Brian Cox and Jeff Forshaw
(De Capo, 2010)

Websites

Eric Weisstein's World of Physics
http://scienceworld.wolfram.com/physics/

Frequently Asked Questions in Physics
http://math.ucr.edu/home/baez/physics/
Maintained by Don Koks.

Official Website of Jim Al-Khalili
http://www.jimal-khalili.com

Official Website of Brian Clegg
http://www.brianclegg.net

Official Website of Frank Close
http://www.frankclose.net/

Thoughts on Life, the Universe, and everything
http://thecuriousastronomer.wordpress.com
Maintained by Rhodri Evans.

Journals/Articles

Do tachyons exist?
http://math.ucr.edu/home/baez/physics/ParticleAndNuclear/tachyons.html

Quantum Entanglement and Information, Stanford Encyclopedia of Philosophy
http://plato.stanford.edu/entries/qt-entangle/

Testing the Multiverse, article on FQXI website by Miriam Frankel
http://fqxi.org/community/articles/display/155

Parallel Universes by Max Tegmark, Scientific American, 2003
http://space.mit.edu/home/tegmark/PDF/multiverse_sciam.pdf

Faster than the speed of light? We'll need to be patient by Jim Al-Khalili
http://www.guardian.co.uk/commentisfree/2011/nov/23/faster-speed-of-light-boxers

In a parallel universe, this theory would make sense by Jim Al-Khalili
http://www.guardian.co.uk/commentisfree/2007/dec/01/comment.spaceexploration

WHAT IF
INDEX

WHAT IF ACKNOWLEDGEMENTS

Picture Credit Acknowledgements

The publisher would like to thank the following individuals and organizations for their kind permission to reproduce the images in this book. Every effort has been made to acknowledge the pictures, and we apologize if there are any inadvertent omissions.

All images from Shutterstock, Inc./ www.shutterstock.com.

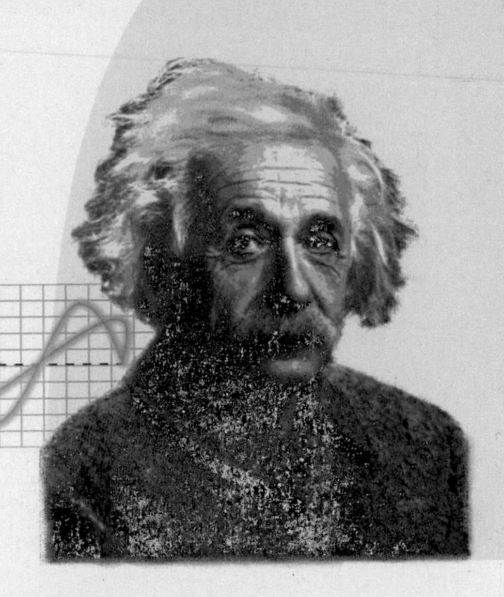